Current Perspectives in Kidney Diseases

Contributions to Nephrology

Vol. 190

Series Editor

Claudio Ronco Vicenza

Current Perspectives in Kidney Diseases

Volume Editors

Gaetano La Manna Bologna
Claudio Ronco Vicenza

28 figures,18 in color, and 15 tables, 2017

KARGER Basel · Freiburg · Paris · London · New York · Chennai · New Delhi · Bangkok · Beijing · Shanghai · Tokyo · Kuala Lumpur · Singapore · Sydney

Contributions to Nephrology
(Founded 1975 by Geoffrey M. Berlyne)

Gaetano La Manna
Nephrology, Dialysis and Renal Transplant Unit
St. Orsola Hospital
University of Bologna
Via Massarenti 9
IT–40138 Bologna (Italy)

Claudio Ronco
Department of Nephrology, Dialysis and Transplantation
International Renal Research Institute of Vicenza (IRRIV)
San Bortolo Hospital
Viale Rodolfi 37
IT–36100 Vicenza (Italy)

Library of Congress Cataloging-in-Publication Data

Names: La Manna, Gaetano, editor. | Ronco, C. (Claudio), 1951- editor.
Title: Current perspectives in kidney diseases / volume editors, Gaetano La Manna, Claudio Ronco.
Description: Basel ; New York : Karger, 2017. | Series: Contributions to nephrology, ISSN 0302-5144 ; v. 190 | Includes bibliographical references and indexes.
Identifiers: LCCN 2017019015 (print) | LCCN 2017018454 (ebook) | ISBN 9783318060614 (e-book) | ISBN 9783318060607 (hard cover : alk. paper) | ISBN 9783318060614 (eISBN)
Subjects: | MESH: Kidney Diseases--therapy | Renal Dialysis | Kidney Transplantation
Classification: LCC RC903 (print) | LCC RC903 (ebook) | NLM WJ 300 | DDC 616.6/1--dc23
LC record available at https://lccn.loc.gov/2017019015

Bibliographic Indices. This publication is listed in bibliographic services, including Current Contents® and Index Medicus.

Disclaimer. The statements, opinions and data contained in this publication are solely those of the individual authors and contributors and not of the publisher and the editor(s). The appearance of advertisements in the book is not a warranty, endorsement, or approval of the products or services advertised or of their effectiveness, quality or safety. The publisher and the editor(s) disclaim responsibility for any injury to persons or property resulting from any ideas, methods, instructions or products referred to in the content or advertisements.

Drug Dosage. The authors and the publisher have exerted every effort to ensure that drug selection and dosage set forth in this text are in accord with current recommendations and practice at the time of publication. However, in view of ongoing research, changes in government regulations, and the constant flow of information relating to drug therapy and drug reactions, the reader is urged to check the package insert for each drug for any change in indications and dosage and for added warnings and precautions. This is particularly important when the recommended agent is a new and/or infrequently employed drug.

www.karger.com
Printed on acid-free and non-aging paper (ISO 9706)
ISSN 0302–5144
e-ISSN 1662–2782
ISBN 978–3–318–06060–7
e-ISBN 978–3–318–06061–4

Contents

Present and Future of Acute and Chronic Kidney Disease

Recent advancements in technology and new discoveries represent a real renaissance in the field of Nephrology. After reaching a milestone in diagnostic techniques and renal replacement therapy, we are now moving towards a new generation of biomarkers for AKI, finding new possibilities of tracking damage progression in chronic nephropathies, testing new dialysis membranes and discovering new approaches for kidney transplantation. In spite of limited financial resources, several innovative opportunities are emerging in many areas of this scientific discipline, with the potential of shedding light on paths that have been either stagnant or unchanged for many years.

In chronic hemodialysis, new information about blood purification techniques and mechanisms of solute removal are now available. Adsorption and specific binding of molecules onto synthetic sorbent particles, the use of high cut-off membranes with novel electrochemical characteristics and expanded spectrum of removal characteristics have improved the ability to clear toxins in the middle-to-high molecular weight, resulting in improvements of clinical outcome.

The innovative approach to fluid management represents a further development of chronic extracorporeal treatment, strengthening the possibilities for home dialysis, with significant improvements in terms of quality of life and rehabilitation of kidney patients.

In acute renal replacement therapies, citrate anticoagulation, high cut-off membrane filters, specifically designed modules, combined techniques, coupled filtration and adsorption seem to permit synergistic interventions in the critical care setting, where extracorporeal therapies represent a method of organ support beyond kidney replacement.

Glomerulopathy therapies are in a phase of evolution. Recent evidences and ongoing trials reveal novel possibilities for treatment, thus enlarging the armamentarium of medications available for nephrologists, with satisfying effectiveness also in the long-term follow-up.

The role and validation of kidney injury biomarkers are important to establish reliable, reproducible, cost-effective, non-invasive early diagnosis. Moreover, the contribution of genetics, besides its diagnostic role, represents an essential predictive tool for risk stratification, disease progression and outcomes, providing the prospect of patient-tailored treatments based on genotype.

In addition, there is a growing body of research on mineral metabolism alterations, commonly identified as a trigger of cardiovascular disease in chronic kidney disease patients, with the involvement of FGF 23 and Klotho, especially in the earlier stages of the disease. The role of these molecules in the regulation of phosphorus metabolism represents only the tip of an iceberg: the impairment of a complex system underlying the progression of renal disease and cardiovascular damage, in a triangle involving heart, kidney and bone, with the latter representing an endocrine organ.

Nowadays, other areas, including the arteriovenous fistula itself, appear not to be vascular anastomosis, but rather an "in vivo" model of vascular damage, with the participation of several mediators and stem cells able to delineate the likelihood of restenosis and the risk of failure. Novel imaging techniques offer possibilities for making a clear diagnosis and for proposing effective remedies.

Finally, in view of the immunological complexity of kidney transplant, one of the biggest challenges is the possibility of influencing the outcomes and long-term allograft survival. The latest therapeutic options offer the possibility of implementing effective preventive measures against graft loss. New tools for preventing chronic allograft injury, the leading cause of transplant failure, are currently being developed: among them, gene expression analysis can help in identifying patients at higher risk, who might benefit from early prophylactic plans, mainly aimed at aggressive treatment of modifiable risk factors, through the control of hypertension, hyperglycemia and hyperlipidemia, CNI minimization, protocol biopsies, addressing drug adherence and performing post-transplant BKV surveillance and DSA (donor-specific antibodies) monitoring.

In a nutshell, the contributions and insights in this volume are meant for the main topics of the current research in Nephrology, and to better understand the most promising opportunities for growth and development in all sectors, ensuring a rationale for optimizing future efforts and promoting solid research.

We hope the content of the book will represent a good opportunity for students and professionals to become acquainted with new areas of research in Nephrology and to better understand why this is an era of renaissance of this discipline.

Gaetano La Manna, Bologna
Claudio Ronco, Vicenza

La Manna G, Ronco C (eds): Current Perspectives in Kidney Diseases.
Contrib Nephrol. Basel, Karger, 2017, vol 190, pp 1–18 (DOI: 10.1159/000468912)

Extracorporeal Treatments in Patients with Acute Kidney Injury and Sepsis

Marita Marengo[a] · Sergio Dellepiane[b] · Vincenzo Cantaluppi[c]

[a]Nephrology and Dialysis Unit, ASL CN1, Cuneo, [b]Nephrology, Dialysis and Kidney Transplantation Unit, Department of Medical Sciences, University of Torino, Torino, and [c]Nephrology and Kidney Transplantation Unit, Department of Translational Medicine, University of Eastern Piedmont "A. Avogadro" (UPO), Novara, Italy

Abstract

Acute kidney injury (AKI) is one of the most common sepsis complications, and AKI development increases the risk of sepsis episodes by affecting host immune competence. The concomitance of these 2 clinical syndromes is associated with an extremely poor prognosis with mortality rates ranging from 50 to 70%. These unacceptable outcomes reflect the poor knowledge of the underlying pathogenic mechanisms and the lack of appropriate diagnostic and therapeutic methodologies as well as of appropriate experimental models. However, in recent years new insights have revolutionized the scientific and clinical approach to sepsis-induced AKI (S-AKI) leading to encouraging results. The aim of this paper is to review the extracorporeal treatment of S-AKI with a focus on the most promising experimental techniques and the underlying molecular mechanisms.

Introduction

Sepsis and acute kidney injury (AKI) are 2 acute clinical syndromes that are tightly connected. From one side, AKI represents one of the most common complications of sepsis and on the other, sepsis risk is significantly increased by AKI development [1]. The concomitance of these 2 clinical syndromes, or sepsis-

M.M. and S.D. equally contributed.

associated AKI (S-AKI), is related to an extremely poor prognosis with 28-day mortality rate ≥50% and a high risk of progression toward chronic kidney disease (CKD) in survivors, with increased healthcare costs [2, 3]. Nowadays, sepsis therapy is based on antibiotic administration and on a complex network of supportive cares that do not target the mechanisms of tissue injury: for this reason, several clinical trials aimed at selectively blocking "the magic bullet" involved in sepsis-associated tissue injury failed [4]. Furthermore, classical renal replacement therapies (RRTs) are not associated with an improvement of renal recovery rate, and have a questioned impact on patient survival in randomized clinical trials with a large number of patients.

However, in last years the scientific and clinical approaches to S-AKI have been deeply revolutionized. The 2001 sepsis definition was focused on the interplay between pathogens and immune cells and the consequent Systemic Inflammatory Response Syndrome (SIRS). Subsequently, large observational studies clearly demonstrated that several septic episodes are not simply related to SIRS and that SIRS detection has a poor clinical utility [5]. Indeed, different tissues implement aberrant responses to infections independently from the immune system. Based on these considerations, in 2016 sepsis was re-defined as "a life-threatening organ dysfunction caused by a dysregulated host response to infection" [5]. Moreover, in the recent years new insights have changed the scientific and clinical approaches to S-AKI, leading to some encouraging results. In particular, "Omic" technologies (i.e. genomics, transcriptomics, proteomics and metabolomics) allowed the development of new biomarkers of outcome and tissue injury. In addition, biotechnology advances in the field of extracorporeal techniques lead to the improvement of treatments aimed to remove the circulating inflammatory mediators involved in the pathogenesis of S-AKI.

The aim of this paper is to review the current extracorporeal therapies available for the treatment of S-AKI, with a particular focus on the most promising techniques known to interfere with the molecular mechanisms involved in renal microvascular derangement and tubular epithelial cell (TEC) injury, key features of S-AKI at cell biology and molecular level.

S-AKI: Novel Pathogenic Mechanisms, Biomarkers of Tissue Injury and Promising Therapeutic Targets

A complex network of pathogenic elements sustains the strong relationship between sepsis and kidney dysfunction. The systemic hemodynamic failure occurring during sepsis has been ascribed for decades as the main or the sole cause of AKI. However, recent experimental and clinical studies clearly demonstrated that the hemodynamic resuscitation of septic patients rarely reverts renal failure and that a significant number of S-AKI episodes are not associated with evident hemodynamic changes [3, 6]. Indeed, AKI may develop in the presence of a normal or even increased renal blood flow, suggesting a dissociation between perfusion and kidney function [7]. On this basis, the mechanisms of kidney injury seem to be related not only to ischemia, but also to other causes that have a toxic and/or immunologic nature. Indeed, an increasing body of evidence demonstrated the pivotal role of harmful circulating mediators (in particular, middle molecules), upregulated during S-AKI and potentially removed by extracorporeal therapies [8, 9]. These detrimental factors may reach the kidney by different ways: (1) some molecules can be freely filtered by glomeruli reaching tubular lumen, thus modulating the biological activities of epithelial cells at this level; (2) the same or other molecules directly act on endothelial cells located in the peritubular capillaries inducing a microvascular derangement that leads to alterations of tubular function at the basolateral compartment. These inflammatory mediators finally lead to bioenergetic alteration, loss of cell polarity, apoptosis, enhanced senescence, and fibroblast differentiation of TECs [10].

Several detrimental molecules are known to be potentially involved in renal cell dysfunction (Figure 1) and are classified in the following categories:

Pathogen-Associated Molecular Patterns (PAMPs). This first family includes molecules produced by pathogens that may have a direct cytotoxic effect or are sensed by tissues as an alarm signal after binding to specific receptors. The most studied PAMP is obviously represented by lipopolysaccharide (LPS) that can directly interact with the Toll-like receptor 4 (TLR-4) on immune cells, kidney resident TECs and endothelial cells. Other highly pathogenic PAMPs include porins, mannose-containing glycoproteins, lipoteichoic acid, flagellin, double-strain RNA and quorum sensing molecules. All these elements are able to alter kidney microcirculation, induce apoptosis and functional alterations of tubular cells and concomitantly modulate the immune response in septic patients [3, 11, 12].

Damage-Associated Molecular Patterns (DAMPs). DAMPs are endogenous molecules released by injured or necrotic cells: RNA, single/double strain DNA, ATP, histones and high-mobility group box 1 (HMGB-1). Also, these molecules activate specific receptors located on the surface of both immune and renal cells (i.e., P2Xr for ATP or TLR-2 for HMGB-1) and have physiological roles in spreading "the alert signal," inducing the recruitment of activated immune cells: indeed, the over-activation of DAMP pathways is a further

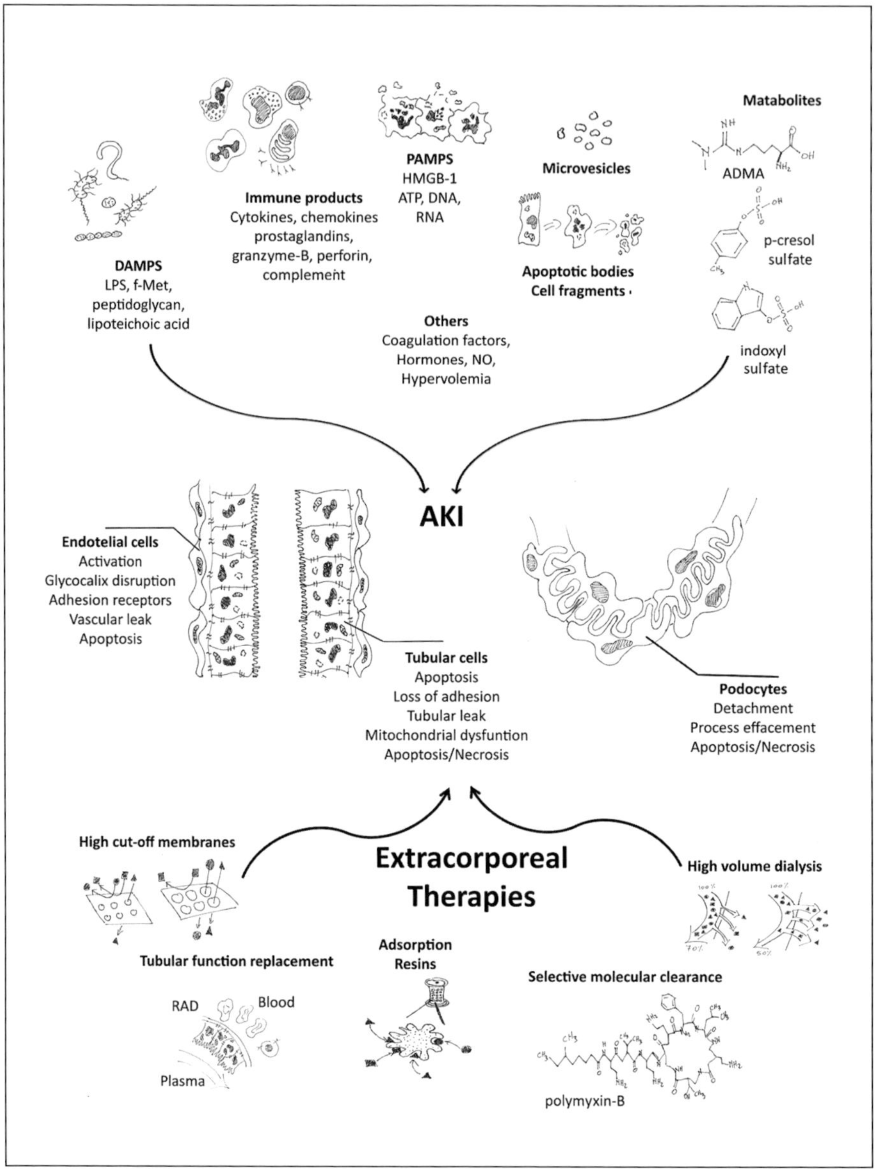

Fig. 1. Modulation of sepsis-associated acute kidney injury by extracorporeal therapies. ADMA, asymmetric dimethylarginine; ATP, adenosine tri-phosphate; DAMPs, damage-associated molecular pathways; HMGB-1, high-mobility group box 1; LPS, lipopolysaccharide; NO, nitric oxide; RAD, renal assist device.

source of renal damage trough direct and indirect (immune-mediated) effects [13, 14].

Inflammatory Cytokines and Chemokines. Cytokines/chemokines are actively produced by injured/activated cells with the aim of modulating the inflammatory response. The immune system is the main source of cytokines, but several other tissues are able to release them. During inflammatory processes, podocytes, TEC and renal endothelium can massively produce IL-18, IL-6 and chemokines such as IL-8 [9, 15]. Several studies have demonstrated the impact of these molecules in S-AKI prognosis; in particular, IL-6 and TNF correlated with AKI severity and with a worse patient survival [16].

Vasoactive Agents and Other Hormones. The release of all the above mentioned molecules coupled with organ dysfunction activates hormones and growth factors involved in homeostasis maintenance. These mediators could only partially counteract tissue injury and promote regeneration processes. However, the excess of signaling of specific stimuli can lead to the development of maladaptive responses. Examples include the massive production of nitric oxide (NO) that induces harmful NO-derived reactive oxygen species, the hyper-activation of the renin-angiotensin-aldosterone system (RAAS) and the sepsis-related catecholamine release that strongly contributes to microvascular vasoconstriction, thrombosis and/or hemorrhage [7, 17].

Immune Products, Activators of Complement and Coagulation Cascades. Cell injury exposes matrix proteins to bloodstream and activates the coagulation cascade. In parallel, complement system is activated by pathogens (mannose pathway), immune complexes (direct pathway) and by downregulation of complement-inhibiting proteins within injured cells (indirect-pathway). Moreover, the loss of glomerular filtration barrier causes proteinuria and exposes complement products to tubular brush border enzymes, thus inducing intra-luminal complement activation [18]. Additionally, activated immune cells release pathogen-killing factors (perforin, granzyme-B, etc.) able to worsen and to perpetuate cell injury [19].

Metabolites and Uremic Toxins. Uremic toxin is an omni-comprehensive term that includes all factors accumulating/upregulated during renal failure that cause any kind of tissue injury. Based on this definition, several molecules included in the previous categories can be defined as uremic toxins (i.e., some cytokines and RAAS). However, the largest part of uremic toxins is constituted by detrimental metabolic products, normally excreted by the kidney. *p*-Cresol sulfate and indoxyl sulfate are protein-bound metabolites not filtered by glomeruli and secreted by TEC; their accumulation in AKI and CKD is associated with several harmful effects such as endothelial injury and immune dysfunction [20].

Extracellular Vesicles (EV), Apoptotic Bodies and Other Cell Fragments. EV are membrane fragments actively produced to shuttle proteins, nucleic acids, lipids and other metabolites from an origin to a target cell [21]. EV play a key role in cell-to-cell communication processes and are classified as exosomes (30–120 nm in size and released by multi-vesicular bodies) or microvesicles (>120 nm and released by a membrane-sorting process), and they are involved in tissue repair and homeostasis. In the course of sepsis, a significant increase of plasma EV concentration is observed [22]. EV can be released by different cell types including monocytes, platelets and injured endothelial cells. The biological effects of EV may change in relation to the state of activation of the origin cell: this is of particular relevance in S-AKI patients in which plasma EV may represent not only a new biomarker for the early detection of disease, but also a key element in the pathogenic mechanisms of renal damage [22, 23]. Plasma EV are able to modulate NO and prostacyclin endothelial release, activate the coagulation cascade and cytokine production. EV isolated from septic animals and injected in healthy ones were able to induce the same functional and biological alterations observed in the course of the systemic inflammatory response, suggesting that EV can somehow transfer the septic disease to a healthy animal [24]. Moreover, sepsis-induced tissue injury promotes the passive release of other cell fragments such as apoptotic and necrotic bodies that have high DAMP concentrations. Several pathogens may also exploit EV to spread the infection or to transport PAMPs/toxins. Of interest, preliminary data from our research group demonstrated that despite their small size, EV are electrically charged and are not easily removed by standard diffusive and/or convective RRT.

Fluid Overload

Recent studies demonstrated the relevant role of fluid overload as AKI determinant. Indeed, a number of retrospective analysis investigating critically ill patients correlated central venous pressure and fluid overload with mortality and worse renal outcomes [25]. All these findings have been recently confirmed by a large multicenter, prospective, observational trial: authors found that the severity and speed of fluid accumulation are independent risk factors for ICU mortality [26].

New Biomarkers and Phenotypic Analysis of Cells in the Peripheral Blood

Diverse studies showed the association between serum levels of specific mediators and a worse outcome in septic patients. Soluble CD40-ligand, Fas-ligand and angiopoietin-2 are middle molecules involved in inflammation, coagulation and apoptotic cell damage that were found to be significantly increased in septic critically ill patients with a worse outcome [27]. In recent years, sepsis research also focused on the alterations of peripheral blood cells: it has been shown that a decreased expression of HLA-DR on monocytes is an indicator of immune paralysis and increased death risk [28]. Similarly, the variation of CD56+ Natural killer T cell count, the increase of CD4+CD25+Foxp3+ T regulatory cells, the decreased number of CD8+ T memory cells and the increased percentage of CD34+CD133+KDR+ endothelial progenitor cells in the peripheral blood have been associated with tissue injury and an increased mortality risk [28].

Extracorporeal Therapies in Patients with Sepsis-Associated AKI

Sepsis and AKI synergistically increase the mortality of ICU patients, and the short- and long-term mortality rates for S-AKI are still unacceptably high (about 50–70%) [29, 30]. Patients with S-AKI have increased mortality compared to non-septic AKI (across all stages of AKI) [31] and to patients with sepsis without AKI. Unfortunately, to date no effective therapy (excluding antimicrobial agents) has been shown to alter the outcome of S-AKI and its management is almost exclusively based on supportive therapies not always able to interfere with the mechanisms of tissue injury or the loss of immune homeostasis.

More than 2 decades ago, it was observed that RRT can remove inflammatory mediators from the plasma of septic patients and improve the pulmonary function [32]. Subsequently, clinical improvements, enhanced cytokine removal and a survival benefit with hemofiltration in septic patients have been reported [33]. Different studies showed that, excluding the absolute indications for RRT (fluid overload, metabolic acidosis, uremia, hyperkalemia and drug intoxication), the decision to initiate dialysis is mainly based on the clinical judgment. Indeed, guidelines regarding timing, indication, modality and dose of RRT for S-AKI patients are still lacking. In this part of the review, we will evaluate the clinical and biological effects of different extracorporeal treatments for S-AKI, focusing on timing, dose and modality of RRT. We will consider new therapeutic strategies based on extracorporeal blood purification techniques that may play a role in improving outcomes and in decreasing progression toward CKD in the near future.

Biological and Clinical Effects of Extracorporeal Treatments

The pathophysiology of sepsis is complex, and renal dysfunction does not simply result from ischemia/hypoperfusion but rather holds a set of inflammation, microcirculatory dysfunction, perfusion deficit, bio-energetic reactions and tubular cell adaptation to injury [34]. The main purpose of blood purification therapies should be to restore homeostasis: as a matter of fact, most sepsis mediators are water-soluble and fall into the "middle-molecular weight" category (about 5–50 kDa) that can be theoretically removed by RRT via convection, diffusion or adsorption. On this basis, we can speculate that RRT can modulate the immune response to sepsis. Furthermore, RRT may have additional clinical benefits including thermal balance, cardiac support achieved by optimization of fluid balance and adequate levels of preload and afterload, protective lung support by the reduction of edema, brain protection with preservation of cerebral perfusion, liver and bone marrow support via the clearance of specific toxins [35].

Timing

Nowadays, there is no consensus on the optimal timing of RRT initiation. Recent studies showed that an excessive delay in RRT start is associated with higher mortality rates and with the worsening of renal function. Retrospective and observational studies suggested that patients with S-AKI treated with early RRT (defined as urea <35.7 mmol/L, start of RRT ≤24 h after the diagnosis of sepsis or by time from ICU admission/initiation of vasopressor infusion) have better survival [36–38]. Also, recent reviews [39, 40] and a meta-analysis [41] showed that earlier start of RRT in critically ill patients with AKI (before the onset of complications and of fluid overload) may have a beneficial impact on survival; however, the same authors emphasized that this conclusion is based on heterogeneous studies that differed in their definition of early and late initiation. Another retrospective review revealed no significant clinical benefit of early RRT initiation in patients presenting with septic shock and AKI [42]. Furthermore, the recent trial from AKIKI group [43] did not show any significant difference in mortality with delayed or early RRT start. The forthcoming Initiation of Dialysis Early versus Late in the Intensive Care Unit study (IDEAL-ICU) will probably help in defining the optimal timing of RRT in S-AKI. Meanwhile, it is acceptable to start RRT at RIFLE injury/failure level as suggested by the Kidney Disease Improving Global Outcomes (KDIGO) guidelines [44]; early initiation of RRT is indicated when fluid overload is excessive or refractory to diuretics [45].

Dose

A relevant issue in RRT for S-AKI is the optimal dose of renal support. The "Vicenza study" conducted by Ronco et al. [46] showed a better survival in patients treated with a filtration rate ≥35 mL/kg/h, in particular, in the presence of sepsis. Subsequent evidence from 2 multicenter trials (RENAL – Randomized Evaluation of Normal versus Augmentated Level Renal Replacement Therapy [47] and ATN – Veterans Affairs/National Institutes of Health Acute Renal Failure Trial Network Study) [48], showed no beneficial effects when high-intensity dose RRT was compared to lower doses. On the contrary, in the RENAL study a post-hoc analysis showed a trend toward a reduction in mortality rate in S-AKI patients treated with high-intensity dose (40 mL/kg/h). Subsequently the IVOIRE study [49] compared high-volume hemofiltration (HVHF, 70 mL/kg/h) to standard-volume hemofiltration (35 mL/kg/h) in septic shock patients without finding a survival or clinical benefit.

Currently, the main aim of S-AKI therapy is early source control and appropriate antibiotic therapy. In many of these "high-volume" studies, no correction was made for antibiotic flux and so, patients may have been underdosed [50]. Additionally, according to the results of the DO-RE-Mi-FA study [26], the actual delivered dose of RRT is approximately 70–90% of the prescription. Thus, prescribing a 25–30 mL/kg/h dose may be more useful in S-AKI, but no strong data support its recommendation. In any case, a particular attention to verify the difference between prescribed versus delivered dose is mandatory.

Modality

Continuous RRT (CRRT) and intermittent hemodialysis (IHD) represent the mainstays of treatment for S-AKI patients requiring RRT. To date, there is no consensus in choosing CRRT or IHD in these patients, but in the presence of hemodynamic instability it is suggested to favor continuous therapies (level 2B in KDIGO guidelines [44]): indeed, hemodynamically unstable patients treated with CRRT remained significantly less dialysis-dependent [51], and CRRT was associated with a trend towards early reduction of vasopressor support [52]. Furthermore, it has been shown that standard IHD has lower capacity to remove several inflammatory cytokines; however, the use of hybrid therapies such as IHD with high cut-off membranes (HCO) or sustained low-efficiency dialysis (SLED) holds some interesting promises. SLED has been shown to provide good tolerability in critically ill patients, excellent clearance of low molecular weight solutes and reasonable clearance of larger molecules able to modulate immune function [53].

Anticoagulation Strategies

The most used anticoagulation strategy is still represented by unfractioned or low molecular weight heparins. However, several problems associated with heparin use should be considered when critically ill patients are subjected to RRT. First, heparin and heparinoids may be contraindicated in particular cases (i.e., heparin-induced thrombocytopenia); moreover, in the case of S-AKI, the low levels of antithrombin III (ATIII) and the putative inflammatory effects of heparin may lead to a premature clotting of the circuit. Intravenous administration of ATIII is recommended to maintain levels higher than 60–70%. In accordance with KDIGO guidelines [44], for patients with hemodynamic instability in which continuous therapies represent the treatment of choice, regional citrate anticoagulation (RCA) should be adopted. Several studies demonstrated the advantage of RCA in comparison to heparin in reducing bleeding and increasing circuit lifespan without clotting episodes. A randomized clinical trial by Oudemans Van Straten et al. [54] also revealed a better survival of RRT patients treated with RCA in comparison to heparin; even though these results have not been replicated in other studies, some specific characteristics of citrate should be considered when S-AKI patients are treated. Indeed, citrate is known to exert potent anti-inflammatory effects such as a reduced formation of platelet-leukocyte complexes and of polymorphonuclear cell degranulation together with the induction of decreased levels of markers of oxidative stress and IL-1-beta. Since citrate may interfere with intracellular calcium signaling that is essential for the release of inflammatory mediators from activated cells, this anticoagulation strategy will be certainly the main actor of future studies aimed at limiting inflammation in the course of S-AKI.

Specific Extracorporeal Therapies for S-AKI

Blood purification therapies able to remove inflammatory substances from the circulation of septic patients include: diffusion-based hemodialysis, convection-based hemofiltration, mixed diffusive-convective therapies (hemodiafiltration), plasma-filtration and adsorption, hemoperfusion and some combinations of them. Nowadays, a consensus on the optimal extracorporeal treatment for S-AKI is still lacking and many randomized clinical trials only reported an improvement of hemodynamics without any positive effect on outcome. Standard RRT therapies using high-flux membranes or adsorptive membranes include continuous veno-venous hemofiltration (CVVH), continuous veno-venous hemodialysis and continuous veno-venous hemodiafiltration have been evaluated for S-AKI treatment. The high-flux membranes have an average cut-off of 30–40 kDa and are able to eliminate significant amounts of inflammatory mediators including chemokines and cytokines. Other types of membranes such as polymethylmethacrylate and AN69ST are capable of adsorbing inflammatory mediators that are known to play a key role in the pathogenic mechanisms of S-AKI; this is the case of HMGB-1 which, despite small size (26 kDa), is not removed by convection but exclusively by adsorption [55]. However, we must emphasize that these inflammatory mediators have a very high generation rate: for this reason, many studies using RRT failed to show any significant modulation of plasma levels of different cytokines [56]. These standard RRT strategies may also adopt the use of HCO, which are porous enough to remove large molecules (approximately 15–60 kDa). Several studies showed clinical benefits associated with the use of HCO membranes: improved immune cell function and removal of inflammatory cytokines, with a decreased dose of norepinephrine in S-AKI patients [57]. An undesired effect of HCO is represented by albumin loss, which can be attenuated by albumin replacement or by using HCO membranes in a diffusive rather than a convective manner [58].

Convection-based high-volume therapies (HVHF) are defined by a flow rate of more than 35 mL/kg/h. To achieve HVHF, it is necessary to use a high permeability membrane with a large surface area and a sieving coefficient close to 1 for a wide spectrum of molecules. HVHF has been shown to improve hemodynamic and survival in patients with refractory septic shock [46]. On the contrary, a recent Cochrane review [59], comparing HVHF with standard dialysis, did not show any improvements in patients' outcome. Furthermore, the use of HVHF may potentially cause increased clearance of antimicrobial agents, electrolyte disturbances and depletion of micronutrients.

Hemoperfusion, hemoadsorption and plasma adsorption are techniques in which a sorbent is placed in an extracorporeal circuit in direct contact with blood or plasma, respectively. Polymyxin B (PMX-B) is a cationic polypeptide antibiotic with known activity directed to gram-negative bacteria and high affinity to endotoxin. The intravenous use of PMX-B is avoided in consideration of its nephrotoxicity and neurotoxicity. PMX-B has been fixed and immobilized onto polystyrene fiber in a hemoperfusion column cartridge that allows LPS removal without the above mentioned toxic effects [36]. The main mechanism of PMX-B action is the removal of circulating endotoxin, although its effects are likely pleiotropic including the aspecific entrapment of inflammatory cells. A preliminary study of PMX-B hemoperfusion added to conventional therapy showed an improvement of hemodynamics, less organ dysfunction and reduced 28-day mortality in patients with severe sepsis or septic shock from abdominal origin [60]. A recent multicenter randomized study did not demonstrate a significant difference in mortality or improvement in organ failure comparing PMX-B hemoperfusion with conventional treatments in patients with septic shock induced by peritonitis [61]. The first randomized, controlled, diagnostic-directed and theragnostic trial named "Evaluating the Use of PMX-B Hemoperfusion in a Randomized controlled trial of Adults Treated for Endotoxemia and Septic shock" (EUPHRATES) is still ongoing in the US and Canada [62]. However, meta-analysis showed that PMX-B treatment is the only adsorption strategy to impact patient's mortality [63]. Moreover, the PMX-B-associated improvement of the SOFA score has been reported and experimental studies showed that after hemoperfusion, the pro-apoptotic effect of septic plasma on cultured human kidney-derived TECs was significantly decreased. These clinical and experimental results suggest that PMX-B hemoperfusion remove from the bloodstream LPS and other potential mediators of S-AKI [64].

Among other emerging adsorption strategies, CytoSorb is a biocompatible polymer, able to aspecifically adsorb inflammatory cytokines from the bloodstream. Kellum et al. [65] demonstrated, in experimental models of sepsis, that this sorbent is able to modulate chemokine gradients between infected tissue and healthy organs, thus directing leukocyte trafficking in the sites of tissue injury [41]. An increasing number of evidences based on clinical cases suggested that CytoSorb is efficient in removing septic mediators, bilirubin and myoglobin.

Coupled plasma filtration adsorption (CPFA) is based on aspecific adsorption of inflammatory mediators onto a sorbent located after a plasma filter: CPFA can be integrated with a standard RRT circuit. Experimental studies showed that CPFA is able to prevent microvascular failure with a consequent improvement of hemodynamic instability, respiratory parameters and a decreased use of vasopressors. The COMPACT study demonstrated that high volumes of plasma exchange using CPFA are mandatory to obtain a clinical benefit and the biological effects of protection against microvascular endothelial cell damage and immunoparalysis [66]. The randomized controlled trial COMPACT-2 is still ongoing in different Italian ICUs.

Last, conventional extracorporeal blood purification techniques remove toxic factors by plasma through filtration, thus mimicking kidney glomerular but not tubular function. Indeed, renal parenchyma contributes to metabolite clearance with tubular secretion (essential for protein-bound molecules) and releases several trophic factors such as erythropoietin and vitamin D. Based on this assumption, Dr. Humes group [67] implemented a standard RRT circuit with a polysulfone filter containing living kidney TECs. This renal assist device (RAD) was at first validated in large animal sepsis models: treated animals maintained reabsorption of K+, $HCO3^-$ and glucose, excretion of ammonia and normal levels of 1,25-OH-vitamin D3 [67]. Moreover, the authors found a significant reduction of inflammatory cytokines only in RAD-treated animals; interestingly, RAD was also able to regulate both the early inflammatory cytokines (e.g., IL-6) and the late-phase cytokines involved in post-septic immune-paralysis (IL-10 and G-CSF) [67]. Encouraged by these results, the same group performed a randomized controlled trial and showed a 50% reduction of 180-day mortality in patients treated with RAD compared to standard CVVH [67]. This trial was prematurely interrupted when the investigators observed a significant reduction of mortality also using a sham cartridge not containing viable tubular cells. Subsequently, they developed the so called selective cytophoretic device (SCD): SCD is similar to RAD but without tubular cells in the second filter. In this case, SCD is able to sequestrate activated leukocytes within the membrane, thus inhibiting the release of harmful mediators. Preliminary studies indicated that SCD reduced mortality and dialysis dependence in S-AKI patients. Interestingly, the outcome improvement was observed only when citrate, and not heparin, was used as an anticoagulant (see paragraph on anticoagulation strategies) [68].

The development of cell therapies associated with RRT may lead to a further improvement of S-AKI. Transplantation of mesenchymal stem cells (MSC) has been shown to reduce mortality and organ failure in experimental models of sepsis. Some reports showed the use of bioreactors that coupled standard RRT with a filter containing viable MSC able to secrete regenerative and immunomodulatory factors [69].

Antibiotic Dosing during RRT

In severe sepsis and septic shock early, appropriate, empiric and broad-spectrum antibiotics are the mainstay of treatment and represent a crucial factor in improving the patient outcome. In septic patients under RRT, the optimization of antibiotic dosing is mandatory but, unfortunately, data to guide dosing in these patients are limited. Patients are at risk of both over- and under-dosing with consequent risk of drug toxicity or treatment failure. When an antibiotic regimen is prescribed in S-AKI patients treated with RRT, several factors have to be considered: pharmacokinetics, patient weight, residual renal function, hepatic function, mode of RRT (membrane and surface area, sieving coefficient, effluent and dialysate rate and blood flow rate), minimum inhibitory concentration, volume overload etc. [70]. Studies that determine the serum antibiotic concentrations are very useful in establishing the correct dosage in critically ill patients, but available data are often based on old RRT modalities resulting in unhelpful/inaccurate dosing recommendations. The application of these older doses in Monte Carlo simulation studies revealed that many of the recommended dosing regimens will never attain pharmacodynamic target [71]. For these reasons, some authors encourage clinicians to prescribe antibiotics, in this vulnerable population, with large loading dose and higher maintenance doses to reach the targets [70, 71].

Conclusions

Sepsis is a serious medical condition frequently associated with the development of multiple organ failure and AKI. The association of sepsis and the loss of renal function determine high incidence of mortality and progression toward CKD. These negative results are possibly due to the lack of human invasive studies (i.e., kidney biopsy) and reliable pathogenic models. However, in recent years, large animal studies and ex vivo human experiments have provided new insights into the pathogenesis of S-AKI. Furthermore, the application of new RRT biotechnologies has opened a new scenario with encouraging clinical data. Despite most of these technologies (i.e., RAD, SCD, polymyxin and CPFA) need to be tested in large phase-3 clinical trials. Some technologies displayed impressive changes in patient mortality (up to 50% for RAD) or were proven to be effective by methanalitic investigations.

References

1 Clark E, Bagshaw SM: Long-term risk of sepsis among survivors of acute kidney injury. Crit Care 2014;18:103.

2 Mehta RL, Bouchard J, Soroko SB, Ikizler TA, Paganini EP, Chertow GM, et al: Sepsis as a cause and consequence of acute kidney injury: program to improve care in acute renal disease. Intensive Care Med 2011;37:241–248.

3 Dellepiane S, Marengo M, Cantaluppi V: Detrimental cross-talk between sepsis and acute kidney injury: new pathogenic mechanisms, early biomarkers and targeted therapies. Crit Care 2016;20:61.

4 Ranieri VM, Thompson BT, Barie PS, Dhainaut JF, Douglas IS, Finfer S, et al: Drotrecogin alfa (activated) in adults with septic shock. N Engl J Med 2012;366:2055–2064.

5 Singer M, Deutschman CS, Seymour CW, Shankar-Hari M, Annane D, Bauer M, et al: The third international consensus definitions for sepsis and septic shock (sepsis-3). JAMA 2016;315:801–810.

6 Murugan R, Karajala-Subramanyam V, Lee M, Yende S, Kong L, Carter M, et al: Acute kidney injury in non-severe pneumonia is associated with an increased immune response and lower survival. Kidney Int 2010; 77:527–535.

7 Di Giantomasso D, Morimatsu H, May CN, Bellomo R: Intrarenal blood flow distribution in hyperdynamic septic shock: effect of norepinephrine. Crit Care Med 2003;31:2509–2513.

8 Mariano F, Cantaluppi V, Stella M, Romanazzi GM, Assenzio B, Cairo M, et al: Circulating plasma factors induce tubular and glomerular alterations in septic burns patients. Crit Care 2008;12:R42.

9 Cantaluppi V, Quercia AD, Dellepiane S, Ferrario S, Camussi G, Biancone L: Interaction between systemic inflammation and renal tubular epithelial cells. Nephrol Dial Transplant 2014;29:2004–2011.

10 Peng ZY, Zhou F, Kellum JA: Cross-species validation of cell cycle arrest markers for acute kidney injury in the rat during sepsis. Intensive Care Med Exp 2016;4:12.

11 Gupta RK, Chhibber S, Harjai K: Quorum sensing signal molecules cause renal tissue inflammation through local cytokine responses in experimental UTI caused by Pseudomonas aeruginosa. Immunobiology 2013; 218:181–185.

12 Opal SM, Laterre PF, Francois B, LaRosa SP, Angus DC, Mira JP, et al: Effect of eritoran, an antagonist of MD2-TLR4, on mortality in patients with severe sepsis: the ACCESS randomized trial. JAMA 2013;309:1154–1162.
13 Chen J, Hartono JR, John R, Bennett M, Zhou XJ, Wang Y, et al: Early interleukin 6 production by leukocytes during ischemic acute kidney injury is regulated by TLR4. Kidney Int 2011;80:504–515.
14 Solini A, Usuelli V, Fiorina P: The dark side of extracellular ATP in kidney diseases. J Am Soc Nephrol 2015;26:1007–1016.
15 Gerritsma JS, Hiemstra PS, Gerritsen AF, Prodjosudjadi W, Verweij CL, Van Es LA, et al: Regulation and production of IL-8 by human proximal tubular epithelial cells in vitro. Clin Exp Immunol 1996;103:289–294.
16 Panacek EA, Marshall JC, Albertson TE, Johnson DH, Johnson S, MacArthur RD, et al: Efficacy and safety of the monoclonal anti-tumor necrosis factor antibody F(ab')2 fragment afelimomab in patients with severe sepsis and elevated interleukin-6 levels. Crit Care Med 2004;32:2173–2182.
17 Husi H, Sanchez-Niño MD, Delles C, Mullen W, Vlahou A, Ortiz A, et al: A combinatorial approach of proteomics and systems biology in unravelling the mechanisms of acute kidney injury (AKI): involvement of NMDA receptor GRIN1 in murine AKI. BMC Syst Biol 2013;7:110.
18 Gaarkeuken H, Siezenga MA, Zuidwijk K, van Kooten C, Rabelink TJ, Daha MR, et al: Complement activation by tubular cells is mediated by properdin binding. Am J Physiol Renal Physiol 2008;295:F1397–F1403.
19 Heng B, Li Y, Shi L, Du X, Lai C, Cheng L, et al: A meta-analysis of the significance of granzyme B and perforin in noninvasive diagnosis of acute rejection after kidney transplantation. Transplantation 2015;99:1477–1486.
20 Sun J, Shannon M, Ando Y, Schnackenberg LK, Khan NA, Portilla D, et al: Serum metabolomic profiles from patients with acute kidney injury: a pilot study. J Chromatogr B Analyt Technol Biomed Life Sci 2012;893–894:107–113.
21 Cantaluppi V, Medica D, Mannari C, Stiaccini G, Figliolini F, Dellepiane S, et al: Endothelial progenitor cell-derived extracellular vesicles protect from complement-mediated mesangial injury in experimental anti-Thy1.1 glomerulonephritis. Nephrol Dial Transplant 2015;30:410–422.
22 Zafrani L, Gerotziafas G, Byrnes C, Hu X, Perez J, Lévi C, et al: Calpastatin controls polymicrobial sepsis by limiting procoagulant microparticle release. Am J Respir Crit Care Med 2012;185:744–755.
23 Meziani F, Delabranche X, Asfar P, Toti F: Bench-to-bedside review: circulating microparticles – a new player in sepsis? Crit Care 2010;14:236.
24 Mastronardi ML, Mostefai HA, Meziani F, Martínez MC, Asfar P, Andriantsitohaina R: Circulating microparticles from septic shock patients exert differential tissue expression of enzymes related to inflammation and oxidative stress. Crit Care Med 2011;39:1739–1748.
25 Legrand M, Dupuis C, Simon C, Gayat E, Mateo J, Lukaszewicz AC, et al: Association between systemic hemodynamics and septic acute kidney injury in critically ill patients: a retrospective observational study. Crit Care 2013;17:R278.
26 Garzotto F, Ostermann M, Martín-Langerwerf D, Sánchez-Sánchez M, Teng J, Robert R, et al: The dose response multicentre investigation on fluid assessment (DoReMIFA) in critically ill patients. Crit Care 2016;20:196.
27 Lorente L, Martín MM, Varo N, Borreguero-León JM, Solé-Violán J, Blanquer J, et al: Association between serum soluble CD40 ligand levels and mortality in patients with severe sepsis. Crit Care 2011;15:R97.
28 Rimmelé T, Payen D, Cantaluppi V, Marshall J, Gomez H, Gomez A, et al: Immune cell phenotype and function in sepsis. Shock 2016;45:282–291.
29 Barbar SD, Binquet C, Monchi M, Bruyère R, Quenot JP: Impact on mortality of the timing of renal replacement therapy in patients with severe acute kidney injury in septic shock: the IDEAL-ICU study (initiation of dialysis early versus delayed in the intensive care unit): study protocol for a randomized controlled trial. Trials 2014;15:270.

30 Bagshaw SM, Uchino S, Bellomo R, Morimatsu H, Morgera S, Schetz M, et al: Septic acute kidney injury in critically ill patients: clinical characteristics and outcomes. Clin J Am Soc Nephrol 2007;2:431–439.
31 Oppert M, Engel C, Brunkhorst FM, Bogatsch H, Reinhart K, Frei U, et al: Acute renal failure in patients with severe sepsis and septic shock – a significant independent risk factor for mortality: results from the German Prevalence Study. Nephrol Dial Transplant 2008;23:904–909.
32 Hoffmann JN, Hartl WH, Deppisch R, Faist E, Jochum M, Inthorn D: Hemofiltration in human sepsis: evidence for elimination of immunomodulatory substances. Kidney Int 1995;48:1563–1570.
33 Bellomo R, Kellum JA, Gandhi CR, Pinsky MR, Ondulik B: The effect of intensive plasma water exchange by hemofiltration on hemodynamics and soluble mediators in canine endotoxemia. Am J Respir Crit Care Med 2000;161:1429–1436.
34 Gomez H, Ince C, De Backer D, Pickkers P, Payen D, Hotchkiss J, et al: A unified theory of sepsis-induced acute kidney injury: inflammation, microcirculatory dysfunction, bioenergetics, and the tubular cell adaptation to injury. Shock 2014;41:3–11.
35 Ronco C, Kellum JA, Bellomo R, House AA: Potential interventions in sepsis-related acute kidney injury. Clin J Am Soc Nephrol 2008;3: 531–544.
36 Oh HJ, Shin DH, Lee MJ, Koo HM, Doh FM, Kim HR, et al: Early initiation of continuous renal replacement therapy improves patient survival in severe progressive septic acute kidney injury. J Crit Care 2012;27:743.e9–e18.
37 Chon GR, Chang JW, Huh JW, Lim CM, Koh Y, Park SK, et al: A comparison of the time from sepsis to inception of continuous renal replacement therapy versus RIFLE criteria in patients with septic acute kidney injury. Shock 2012;38:30–36.
38 Carl DE, Grossman C, Behnke M, Sessler CN, Gehr TW: Effect of timing of dialysis on mortality in critically ill, septic patients with acute renal failure. Hemodial Int 2010;14:11–17.
39 Ostermann M, Dickie H, Barrett NA: Renal replacement therapy in critically ill patients with acute kidney injury – when to start. Nephrol Dial Transplant 2012;27:2242–2248.
40 Joannidis M, Forni LG: Clinical review: timing of renal replacement therapy. Crit Care 2011;15:223.
41 Karvellas CJ, Farhat MR, Sajjad I, Mogensen SS, Leung AA, Wald R, et al: A comparison of early versus late initiation of renal replacement therapy in critically ill patients with acute kidney injury: a systematic review and meta-analysis. Crit Care 2011;15:R72.
42 Shum HP, Chan KC: Timing for initiation of continuous renal replacement therapy in patients with septic shock and acute kidney injury. Ther Apher Dial 2013;17:643.
43 Gaudry S, Hajage D, Schortgen F, Martin-Lefevre L, Pons B, Boulet E, et al: Initiation strategies for renal-replacement therapy in the intensive care unit. N Engl J Med 2016; 375:122–133.
44 Kellum JA, Lameire N; KDIGO AKI Guideline Work Group: Diagnosis, evaluation, and management of acute kidney injury: a KDIGO summary (part 1). Crit Care 2013;17: 204.
45 Fülöp T, Pathak MB, Schmidt DW, Lengvárszky Z, Juncos JP, Lebrun CJ, et al: Volume-related weight gain and subsequent mortality in acute renal failure patients treated with continuous renal replacement therapy. ASAIO J 2010;56:333–337.
46 Ronco C, Bellomo R, Homel P, Brendolan A, Dan M, Piccinni P, et al: Effects of different doses in continuous veno-venous haemofiltration on outcomes of acute renal failure: a prospective randomised trial. Lancet 2000; 356:26–30.
47 RENAL Replacement Therapy Study Investigators, Bellomo R, Cass A, Cole L, Finfer S, Gallagher M, et al: Intensity of continuous renal-replacement therapy in critically ill patients. N Engl J Med 2009;361:1627–1638.
48 VA/NIH Acute Renal Failure Trial Network, Palevsky PM, Zhang JH, O'Connor TZ, Chertow GM, Crowley ST, et al: Intensity of renal support in critically ill patients with acute kidney injury. N Engl J Med 2008;359: 7–20.

49 Joannes-Boyau O, Honoré PM, Perez P, Bagshaw SM, Grand H, Canivet JL, et al: High-volume versus standard-volume haemofiltration for septic shock patients with acute kidney injury (IVOIRE study): a multicentre randomized controlled trial. Intensive Care Med 2013;39:1535–1546.
50 Doyle JF, Forni LG: Update on sepsis-associated acute kidney injury: emerging targeted therapies. Biologics 2016;10:149–156.
51 Prowle JR, Bellomo R: Continuous renal replacement therapy: recent advances and future research. Nat Rev Nephrol 2010;6:521–529.
52 Wald R, Friedrich JO, Bagshaw SM, Burns KE, Garg AX, Hladunewich MA, et al: Optimal mode of clearance in critically ill patients with acute kidney injury (OMAKI) – a pilot randomized controlled trial of hemofiltration versus hemodialysis: a Canadian Critical Care Trials Group project. Crit Care 2012;16:R205.
53 Fiaccadori E, Maggiore U, Parenti E, Giacosa R, Picetti E, Rotelli C, et al: Sustained low-efficiency dialysis (SLED) with prostacyclin in critically ill patients with acute renal failure. Nephrol Dial Transplant 2007;22:529–537.
54 Oudemans-van Straaten HM, Bosman RJ, Koopmans M, van der Voort PH, Wester JP, van der Spoel JI, et al: Citrate anticoagulation for continuous venovenous hemofiltration. Crit Care Med 2009;37:545–552.
55 Yumoto M, Nishida O, Moriyama K, Shimomura Y, Nakamura T, Kuriyama N, et al: In vitro evaluation of high mobility group box 1 protein removal with various membranes for continuous hemofiltration. Ther Apher Dial 2011;15:385–393.
56 Honore PM, Jacobs R, Joannes-Boyau O, De Regt J, De Waele E, van Gorp V, et al: Newly designed CRRT membranes for sepsis and SIRS – a pragmatic approach for bedside intensivists summarizing the more recent advances: a systematic structured review. ASAIO J 2013;59:99–106.
57 Morgera S, Haase M, Kuss T, Vargas-Hein O, Zuckermann-Becker H, Melzer C, et al: Pilot study on the effects of high cutoff hemofiltration on the need for norepinephrine in septic patients with acute renal failure. Crit Care Med 2006;34:2099–2104.
58 Haase M, Bellomo R, Baldwin I, Haase-Fielitz A, Fealy N, Davenport P, et al: Hemodialysis membrane with a high-molecular-weight cutoff and cytokine levels in sepsis complicated by acute renal failure: a phase 1 randomized trial. Am J Kidney Dis 2007;50:296–304.
59 Borthwick EM, Hill CJ, Rabindranath KS, Maxwell AP, McAuley DF, Blackwood B: High-volume haemofiltration for sepsis in adults. Cochrane Database Syst Rev 2017;1:CD008075.
60 Cruz DN, Antonelli M, Fumagalli R, Foltran F, Brienza N, Donati A, et al: Early use of polymyxin B hemoperfusion in abdominal septic shock: the EUPHAS randomized controlled trial. JAMA 2009;301:2445–2452.
61 Payen DM, Guilhot J, Launey Y, Lukaszewicz AC, Kaaki M, Veber B, et al: Early use of polymyxin B hemoperfusion in patients with septic shock due to peritonitis: a multicenter randomized control trial. Intensive Care Med 2015;41:975–984.
62 Klein DJ, Foster D, Schorr CA, Kazempour K, Walker PM, Dellinger RP: The EUPHRATES trial (evaluating the use of polymyxin B hemoperfusion in a randomized controlled trial of adults treated for endotoxemia and septic shock): study protocol for a randomized controlled trial. Trials 2014;15:218.
63 Cruz DN, de Cal M, Piccinni P, Ronco C: Polymyxin-B hemoperfusion and endotoxin removal: lessons from a review of the literature. Contrib Nephrol 2010;167:77–82.
64 Cantaluppi V, Quercia AD, Dellepiane S, Figliolini F, Medica D, De Lena M: [New mechanisms and recent insights in the pathogenesis of acute kidney injury (AKI)]. G Ital Nefrol 2012;29:535–547.
65 Kellum JA, Venkataraman R, Powner D, Elder M, Hergenroeder G, Carter M: Feasibility study of cytokine removal by hemoadsorption in brain-dead humans. Crit Care Med 2008;36:268–272.
66 Livigni S, Bertolini G, Rossi C, Ferrari F, Giardino M, Pozzato M, et al: Efficacy of coupled plasma filtration adsorption (CPFA) in patients with septic shock: A multicenter randomised controlled clinical trial. BMJ Open 2014;4:e003536.
67 Tumlin J, Wali R, Williams W, Murray P, Tolwani AJ, Vinnikova AK, et al: Efficacy and safety of renal tubule cell therapy for acute renal failure. J Am Soc Nephrol 2008;19:1034–1040.

68 Tumlin JA, Chawla L, Tolwani AJ, Mehta R, Dillon J, Finkel KW, et al: The effect of the selective cytopheretic device on acute kidney injury outcomes in the intensive care unit: a multicenter pilot study. Semin Dial 2013;26: 616–623.
69 Tsai AC, Ma T: Expansion of human mesenchymal stem cells in a microcarrier bioreactor. Methods Mol Biol 2016;1502:77–86.
70 González de Molina FJ, Ferrer R: Appropriate antibiotic dosing in severe sepsis and acute renal failure: factors to consider. Crit Care 2011;15:175.
71 Shaw AR, Chaijamorn W, Mueller BA: We underdose antibiotics in patients on CRRT. Semin Dial 2016;29:278–280.

Prof. Vincenzo Cantaluppi
SCDU Nefrologia e Trapianto Renale, Dipartimento di Medicina Traslazionale
Università del Piemonte Orientale (UPO) "A. Avogadro"
Azienda Ospedaliera Universitaria "Maggiore della Carità"
Corso Mazzini 18, IT–28100 Novara (Italy)
E-Mail vincenzo.cantaluppi@med.uniupo.it

La Manna G, Ronco C (eds): Current Perspectives in Kidney Diseases.
Contrib Nephrol. Basel, Karger, 2017, vol 190, pp 19–30 (DOI: 10.1159/000468833)

Citrate Anticoagulation during Continuous Renal Replacement Therapy

Davide Ricci · Laura Panicali · Maria Grazia Facchini · Elena Mancini

Nephrology Dialysis Hypertension Department, Teaching Hospital Policlinico S. Orsola-Malpighi, Bologna, Italy

Abstract

During extracorporeal dialysis, some anticoagulation strategy is necessary to prevent the coagulation of blood. Heparin has historically been used as an anticoagulant because of its efficacy combined with low cost. However, a variable incidence of hemorrhagic complications (5–30%) has been documented in patients undergoing continuous renal replacement therapy (CRRT) with heparin as an anticoagulant. Citrate has anticoagulation properties secondary to its ability to chelate calcium, which is necessary for the coagulation cascade. Citrate may thus be used in a regional anticoagulation (RCA), limited to the extracorporeal circuit of CRRT, to avoid systemic anticoagulation. Recent meta-analysis confirmed the advantage of RCA over heparin in terms of incidence of bleeding during CRRT. Moreover, an increase in filter lifespan is documented, with a secondary advantage in reaching the prescribed dialysis dose. In our experience, we could confirm this positive effect. In fact, with a progressive increase in the proportion of CRRT with citrate as RCA, we obtained a reduction in the number of filters used for every 72 h of treatment (from 2.4 in 2011 to 1.3 in 2015), and most importantly, a reduction in the difference between the prescribed and delivered dialysis doses (from 22 to 7%). Citrate has an intense effect on the acid-base balance as well, if fully metabolized through the Krebs cycle, due to the production of bicarbonate. Even more severely ill patients, such as those with liver dysfunction, may be treated with RCA without severe complications, because modern machines for CRRT are equipped with simple systems that are able to manage the citrate infusion and control the calcium levels, with minimal risks of metabolic derangements.

Introduction

Continuous renal replacement therapy (CRRT) is widely used for treating acute kidney injury (AKI). Although there are several experiences on CRRT without anticoagulation [1], some anticoagulation strategy is generally thought to be required to avoid circuit blood clotting. During CRRT, blood is conducted through an extracorporeal circuit, activating coagulation by a complex interaction between patient and circuit. Critically ill patients may also develop a procoagulant state frequently due to sepsis; activation of the coagulation system is triggered by proinflammatory cytokines that enhance the expression of tissue factor on activated mononuclear and endothelial cells and simultaneously downregulate natural anticoagulants, thus initiating thrombin generation, subsequent activation of platelets, and inhibition of fibrinolysis. The beginning of clotting in the extracorporeal circuit has traditionally been attributed to contact activation of the intrinsic coagulation system. However, the bioincompatibility reaction is more complex and is incompletely understood. The activation of tissue factor, leucocytes, and platelets plays an additional role [2].

In spite of the high bleeding risk related to AKI itself [3], systemic anticoagulation with unfractioned heparin has historically been used to maintain the patency of the extracorporeal circuit for CRRT. However, a high incidence of hemorrhagic complications has been documented with wide variability (5–30%) related to differences in patient populations and anticoagulation protocols [4]. Furthermore, a kind of "inverse correlation" between the hemorrhagic risk for the patient and the coagulation risk for the circuit was established in 1996 by Wetering. The higher the attention given to avoid the risk of bleeding (by reducing heparin), the higher is the incidence of thrombotic events to the extracorporeal circuit, and vice versa [5].

To avoid this drawback, both prostaglandin E1 and prostaglandin I2 have been tested in CRRT, but their high cost and hypotension due to vasodilatation are distinct limits to their routine use [2]. Regional citrate anticoagulation (RCA) offers an attractive alternative, and the 2012 Kidney Disease Improving Global Outcomes Clinical Practice Guidelines for AKI recommend the use of RCA as the preferred anticoagulation modality for CRRT in patients without contraindications for citrate, even in the absence of increased bleeding risk or impaired coagulation [6].

Citrate: Mechanism of the Anticoagulant Action

Sodium citrate, infused before the filter, chelates calcium, essential to the coagulation process. The ensuing regional hypocalcemia in the filter inhibits thrombin generation. Citrate is partially removed by filtration or dialysis [7], and the re-

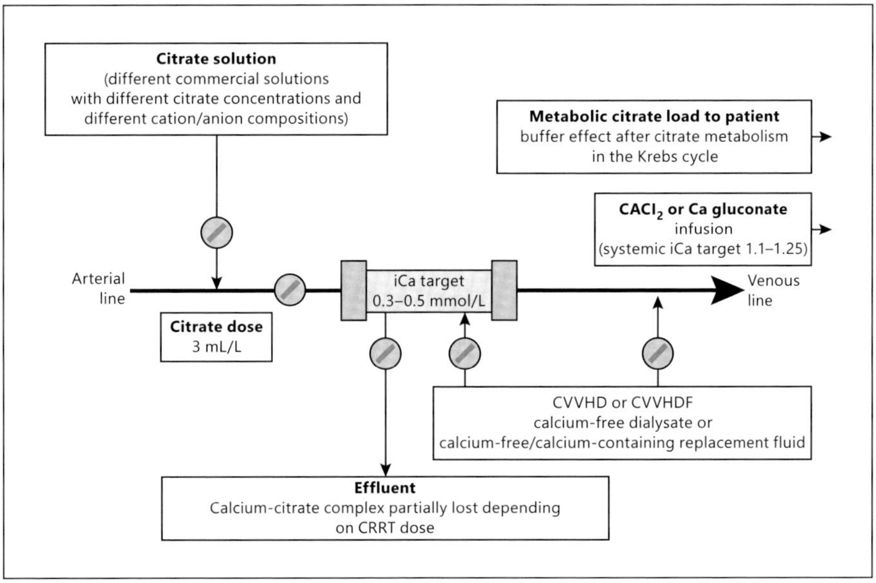

Fig. 1. Schematic representation of the regional citrate anticoagulation implemented in an extracorporeal circuit of CRRT. The citrate solution is infused before the dialyzer, at a flow rate proportional to blood flow, citrate dose, and solution citrate concentration. Citrate chelates calcium: the resulting effect is the extreme reduction of ionized calcium (iCa) in blood (ideal target 0.3–0.4 mmol/L). Calcium must be administered to the patient to restore the physiological level (1.1–1.25 mmol/L). A specific infusion line is needed either downstream of the dialyzer or directly to the patient. The calcium-citrate complexes are partially lost in the effluent, depending on the dialysis dose. The metabolic load of citrate to the patient will be the difference between citrate infused into the circuit and citrate lost with the effluent (correlation between the effluent volume and the amount of citrate lost).

maining amount, infused into the patient, is rapidly metabolized in the citric acid (Krebs) cycle, especially in the liver, muscle, and renal cortex. The calcium previously chelated is released into the patient's blood as a result of citrate metabolism, whereas the calcium lost with the effluent needs to be replaced by calcium solution administered to the patient. Systemic coagulation is thus unaffected (Fig. 1).

For anticoagulation, the citrate dose is adjusted to blood flow so as to achieve an ionized calcium (iCa) concentration <0.3–0.5 mmol/L in the filter [4]; the lower the calcium concentration, the higher the degree of anticoagulation [8]. Some protocols use a fixed dose of citrate in relation to blood flow according to an algorithm, with the target of about 3 mmol citrate/L blood flow. Other protocols adjust the citrate dose by measuring the post-filter iCa, which complicates the intervention but optimizes anticoagulation [2, 4].

RCA versus Heparin: Incidence of Bleeding

Two recent meta-analyses [9, 10] confirmed the advantage of RCA over heparin in terms of incidence of bleeding during CRRT.

In particular, all studies considered by Wu et al. [9] assessed the incidence of bleeding, and a significant difference was found between the 2 groups (RCA vs. heparin), with fewer patients in the citrate group suffering from major bleeding (RR 0.34; 95% CI 0.17–0.65). The number of people needed to receive the treatment before one person would experience a beneficial outcome (number needed to treat) was 6.87.

Again, in a small study that randomized 48 patients for continuous venovenous hemofiltration (CVVH) with RCA or systemic heparin anticoagulation, Betjes observed a significantly higher incidence of bleeding and highlighted that the mean need for red blood cell transfusion was more than double in patients receiving unfractioned heparin (0.88 vs. 0.43 units of packed red blood cell per day of CVVH) [11]. More recently, in a larger study, Morabito et al. [12] likewise showed a lower transfusion rate during RCA-CVVH than with heparin (0.29 vs. 0.62, $p = 0.017$) or no-AC (0.29 vs. 0.64 blood units/day, $p = 0.019$), probably in part due to the significantly lower need for filter set replacement during RCA-CVVH. In addition, RCA was associated with an increase in the platelet count and antithrombin-III activity, thus avoiding platelet concentrate administration and antithrombin-III supplementation.

If the use of citrate is associated with a lower hemorrhagic risk (compared to heparin) and a reduction in blood loss for technical reasons (lower number of filters replaced), it would be reasonable to expect an indirect positive effect on the survival as well, but no strong evidence is presently available about any such possible favorable effects of RCA. Only in one large single-center randomized trial, including 200 critically ill patients on CRRT using nadroparin or citrate anticoagulation, was RCA associated with a surprising 15% absolute increase in 3-month survival, which was not fully clarified by the lower incidence of bleeding. Moreover, post-hoc analysis revealed that RCA might be particularly advantageous in specific clinical conditions (e.g., surgery, sepsis, severe multiple organ dysfunction syndrome, and younger age). To explain this survival advantage of RCA, the authors called into question the possible positive effect of citrate in the "inflammation network" (less polymorphonuclear and platelet degranulation due to hypocalcemia inside the filter, more substrate availability for citric acid cycle maintaining redox state) [13].

Two subsequent multicenter randomized trials, as large as the previous one, comparing unfractioned heparin with RCA in 170 [14] and 212 [15] patients undergoing CRRT, disconfirmed the RCA survival benefit. Finally, the recent meta-analysis by Liu et al. [10] clarified that there was no significant difference in mortality between the citrate and heparin groups.

RCA and Circuit Life Span

The circuit life span is affected by many factors, such as the patient's clinical condition, coagulation status, patency of vascular access, modality of CRRT, and filtration fraction. The various confounding factors may cause a high heterogeneity among trials but the recent meta-analysis by Liu et al. [10] suggested that the choice of anticoagulant between heparin and citrate may play a pivotal role. In particular, RCA may have an advantage in prolonging the circuit life span. Thirteen trials that investigated the circuit life span of citrate versus heparin groups during CRRT were taken into consideration. The circuit duration before clotting was significantly longer in the citrate group than in the heparin group, with a mean difference (MD) of 15.69 h (95% CI 9.30–22.08, $p < 0.01$). Due to the remarkable heterogeneity mentioned above, subgroup analyses were performed by the authors studying CVVH, CVVHDF, pre-dilution, and post-dilution groups separately. Overall, in the CVVH (MD 8.18, 95% CI 3.86–12.51, $p < 0.01$) and pre-dilution subgroups (MD 17.51, 95% CI 9.85–25.17, $p < 0.01$), the circuit life span was significantly longer in the citrate group than in the heparin group [10].

In our experience (we started using citrate in 2010), in the 5 years from 2011 to 2015, we performed a total of 7,316 twenty four hour-treatments (CRRT) with a progressively increasing proportion of treatments using RCA (from 2.9% in 2011 to 49.5% in 2015). We observed a progressive reduction in the number of filters used for every 72 h of treatment (from 2.4 in 2011 to 1.3 in 2015). Although we did not directly collect these data, because the analysis is retrospective, we are in agreement with the reports by authors (Schilder et al. [18]) that one can assume a longer filter duration.

In any case, the circuit life span cannot be considered a positive aim per se, since we should not forget that membrane depurative performances reduce over time. For example, in 2 small trials [16, 17] the authors obtained a median circuit lifetime of 70 and 124 h, respectively during CRRT with RCA. Even though, in the first of these 2 studies, Monchi et al. [16] noted that after 96 h the sieving coefficient for β2-microglobulin was higher than 0.8, it is commonly accepted that such a long filter life span is not the target. What happens in the first 48–72 h is more interesting to understand.

Schilder et al. [18] showed that the down-time (the time when the treatment does not work) within 72 h was less with citrate (1 h for RCA vs. 3 h for heparin,

$p = 0.002$), as were the number of filters used (1 for RCA vs. 2 for heparin, $p = 0.002$). There was a higher incidence of circuit disconnection due to clotting of the circuit in the heparin group (51 vs. 24% in the citrate group) and more elective filter changes in the citrate group (30 vs. 9% in the heparin group, $p = 0.01$). Both Morabito et al. [12] and Stucker et al. [19] showed from similar data that increased filter lifespan with RCA means less treatment interruption and more effective dialysis. These authors confirmed that, starting from the same prescribed dose, the effective delivered daily CRRT dose was higher in the RCA group than in the heparin group in accordance with the well-known concept that the so-called "down-time" is one of the most important determinants of good correspondence between prescribed and delivered doses in CRRT [20].

Our data are consistent with these observations. In fact, in increasing the percentage of treatment using citrate as an anticoagulant (in 2011: no anticoagulation 59.1%, heparin 37.8%, RCA 2.9% of CRRT; in 2015: no anticoagulation 27%, heparin 23.4%, RCA 49.5% of CRRT), the discrepancy between the prescribed and delivered dialysis doses decreased from 22.3% in 2011 to 7% in 2016 (Fig. 2; personal data, not published).

Citrate: Anticoagulant and Buffer

Even though citrate is primarily used for extracorporeal anticoagulation, it has a significant effect on the acid-base balance as well. Anticoagulant and acid-base effects are not directly related. The degree of anticoagulation depends on the citrate dose and hypocalcemia (in the extracorporeal circuit), while the effect on the acid-base status depends on citrate metabolism.

The citrate metabolic load to the patient is the difference between the citrate infused into the CRRT circuit and the quantity of citrate lost in the effluent. In fact there is a direct positive correlation between the effluent volume and the amount of citrate lost [7]. With the more commonly reported citrate protocols, the citrate load is approximately 10–20 mmol/h. This citrate load to the patient is quickly metabolized through the aerobic pathways of the Krebs cycle in the liver, skeletal muscle, and kidney. For each 1 mmol citrate metabolized in the Krebs cycle, 3 mmol hydrogen ions are consumed and 3 mmol bicarbonate is generated, assuming that the citrate is completely metabolized. The resulting bicarbonate produced from citrate metabolism along with bicarbonate in replacement/dialysis fluids provides the buffer supply to the patient [8].

But the buffer power of a citrate solution also depends on the proportion of strong cations in the fluid counterbalancing the citrate anion. Generally, citrate

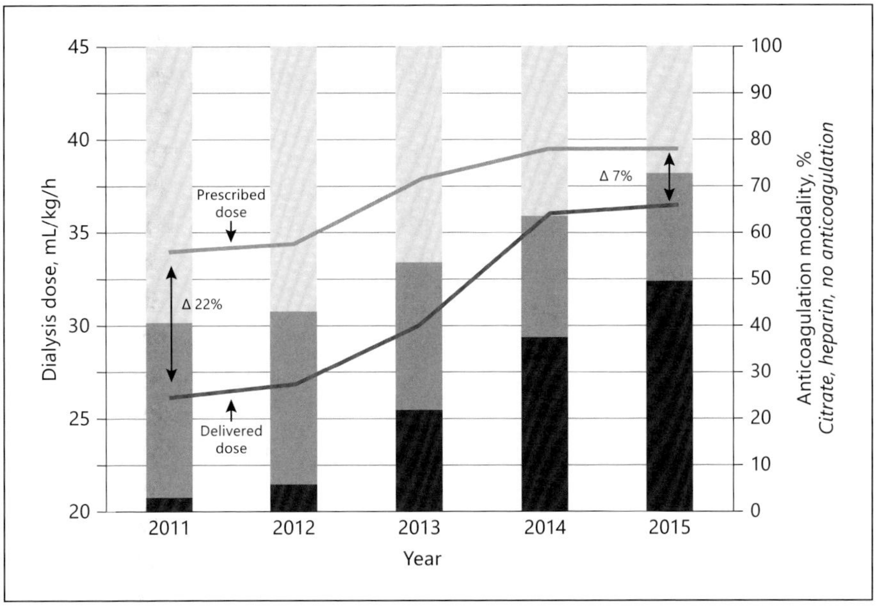

Fig. 2. Change in the CRRT anticoagulation modality from 2011 to 2015 in our Nephrological Department and relative change in the difference between the prescribed and delivered dialysis doses. Columns represent the ratio of different anticoagulation modalities per each year (light grey stands for no anticoagulation; medium grey for heparin; dark grey for citrate). Over the years, we progressively increased the number of treatments with regional citrate anticoagulation modality, which, in 2015, reached near half of the overall number of CRRT performed in the intensive care units. Thanks to the reduction of the down-time due to filter clotting, the discrepancy between the prescribed and delivered dialysis doses (arrows) could be progressively reduced, from 22% in 2011 to 7% in 2015. For the same reason, we could effectively increase the prescribed dose.

is available at various concentrations of trisodium salt, but in some commercial solutions hydrogen is used (citric acid) instead of sodium. Hydrogen does not act as a buffer. The Stewart approach to the acid-base equilibrium [21] provides an interesting tool for a deeper comprehension of the buffering effect of citrate. After metabolism of citrate, the remaining sodium increases the strong ion difference (SID) in body fluids: SID = $(Na^+ + K^+ + Ca^{2+} + Mg^{2+}) - (Cl^- + lactate^-)$. An increased SID produces alkalosis, while the infusion of a zero-SID fluid (such as saline) decreases the SID of body fluids and causes acidosis. As the citrate anion contrasts the cation load, the citrate solution has per se a zero SID (SID = $[Na^+ + K^+ + Ca^{2+} + Mg^{2+}] - [Cl^- + citrate^{3-}]$) and is therefore potentially acidifying. As a result of citrate metabolism, there appears an alkalizing effect due to the strong cation load.

Indeed, in conditions where citrate metabolism is reduced (severe liver failure, severe tissue hypoxia/hypoperfusion) even trisodium citrate can first cause some acidosis as a zero SID solution, but if citrate is metabolized, free strong cations produce their alkalinizing effect.

So, by the same anticoagulant effect, commercial solutions containing citric acid at least in part will have a lower alkalizing effect, and need to be taken into account in the prescription. In addition, the Stewart approach to acid-base equilibrium, is able to offer an explanation of the fact that different citrate formulations, although infused at similar flow rates and thus giving the patient the same citrate load, may exhibit divergent capacity and speed in correcting acidosis because of a substantial difference in solution SID, and in particular in the chloride content [22].

Metabolic and Electrolyte Disarrangements Due to Citrate Anticoagulation

Despite the reported problems that citrate may induce hypernatremia or hyponatremia, hypercalcemia or hypocalcemia, hypermagnesemia or hypomagnesemia, these complications are quite uncommon when there is strict adherence to the RCA protocols. In particular, hypernatremia is an infrequently observed complication associated with the use of hypertonic solutions without low-sodium concentration dialysate and/or replacement fluids. Calcium and magnesium imbalances might be caused by effluent losses in the form of citrate complexes not adequately corrected by systemic supplementation [8].

Impaired liver function, arterial hypoxia, and reduced tissue perfusion are described in the literature as risk factors for citrate accumulation [23]. Citrate is an intermediate of energy metabolism and is not toxic, but accumulation of it with consecutive reduction of iCa could decrease the cardiac contractility or cause arrhythmias, and symptoms of systemic ionized hypocalcemia. However, there is increasing evidence that at least impaired liver function need not be considered as an absolute contraindication for RCA [24]. Several studies have reported that RCA can be safely used even in this population [25–27].

Monitoring Citrate Anticoagulation during CRRT

Measurement of citrate concentration in the blood is not available in the daily routine, so the most commonly accepted clinical markers for citrate accumulation [4, 8] are as follows:

- An increased ratio of total calcium (tCa) to iCa (tCa/iCa),
- Metabolic acidosis with or without an increased anion gap, and
- Elevated demand for calcium substitution.

A calcium ratio (tCa/iCa) of 2.5 is often indicated as the critical threshold for enhanced risk of metabolic complications due to impaired citrate metabolism. Nevertheless, it has been reported that a lower cutoff of 2.1 appears to accurately predict the citrate accumulation (systemic citrate concentration 1 mmol/L) with high sensitivity and specificity (89 and 100%, respectively) [28].

However, the incidence of metabolic disarrangements resulting from citrate accumulation has been found to be rather low since it generally affects less than 3% of all CRRT patients on RCA, and clinical diagnosis of citrate accumulation is found exclusively in severely ill patients with multiorgan failure and severe lactic acidosis [24].

RCA in Patients with Liver Failure as a Safety Paradigm

The use of RCA in patients with liver dysfunction is often considered hazardous due to deranged liver metabolism and increased risk of citrate accumulation. However, many of the potential risks related to the use of citrate in these patients have been overcome, thanks to the recent evolution in dialysis machine engineering technology [24]. The new software is in fact able to adapt citrate infusion to blood flow changes, thus limiting the risk of an inappropriate citrate/blood flow ratio. Moreover, with CRRT monitors the citrate dose can be modified at any time during the treatment in the event of a documented or suspected citrate overload. Last, modulation of the convective and/or diffusive CRRT dose may prevent the development of citrate accumulation, due to the substantial removal of citrate with the effluent fluid [7].

Recent data on RCA in patients with liver failure undergoing CRRT have provided important information. In spite of substantial increases in citrate levels, the metabolic consequences were less significant than expected: a trend towards metabolic alkalosis was found instead of acidosis, and no significant electrolyte disturbances (including hypocalcemia) were observed [26]. Blood citrate levels were related to the calcium ratio; prothrombin activity <26% and lactate level >3.4 mmol/L proved to be acceptable predictors of citrate accumulation. However, a more recent retrospective study suggested that we may further restrict the exclusion criteria for RCA, describing the association between the incidence of citrate accumulation and more elevated lactate levels (>7.5 mmol/L) [23]. Most of these issues can be deemed valid in the case of extracorporeal liver support treatments usually adopted as a "bridge" to liver transplantation or liver recovery, such as the Molecular Adsorbent Recirculating System and Prometheus System. In our experience with Prometheus, RCA was a safe and effective anticoagulation modality [27, 29, 30].

Conclusion

The use of citrate represents a valid alternative to traditional anticoagulation with heparin during CRRT since it reduces the bleeding risk and increases the efficiency of the treatment. The increase of filter life span in fact, is not a goal to be pursued in itself, but the reduction of the filter clotting events within the first 48–72 h reduces the discrepancy between the prescribed and delivered dialysis doses. With recent technological innovations, the RCT during CRRT can be conducted safely with minimal risk of complications even in patients at the highest risk of citrate accumulation such as patients with liver dysfunction.

References

1 Uchino S, Fealy N, Baldwin I, Morimatsu H, Bellomo R: Continuous venovenous hemofiltration without anticoagulation. ASAIO J 2004;50:76–80.

2 Joannidis M, Oudemans-van Straaten HM: Clinical review: patency of the circuit in continuous renal replacement therapy. Crit Care 2007;11:218.

3 Fiaccadori E, Maggiore U, Clima B, Melfa L, Rotelli C, Borghetti A: Incidence, risk factors, and prognosis of gastrointestinal hemorrhage complicating acute renal failure. Kidney Int 2001;59:1510–1519.

4 Morabito S, Pistolesi V, Tritapepe L, Fiaccadori E: Regional citrate anticoagulation for RRTs in critically ill patients with AKI. Clin J Am Soc Nephrol 2014;9:2173–2188.

5 van de Wetering J, Westendorp RG, van der Hoeven JG, Stolk B, Feuth JD, Chang PC: Heparin use in continuous renal replacement procedures: the struggle between filter coagulation and patient hemorrhage. J Am Soc Nephrol 1996;7:145–150.

6 Kidney Disease: Improving Global Outcomes (KDIGO) Acute Kidney Injury Work Group: KDIGO clinical practice guideline for acute kidney injury. Kidney Int 2012;2(suppl):1–138.

7 Mariano F, Morselli M, Bergamo D, Hollo Z, Scella S, Maio M, Tetta C, Dellavalle A, Stella M, Triolo G: Blood and ultrafiltrate dosage of citrate as a useful and routine tool during continuous venovenous haemodiafiltration in septic shock patients. Nephrol Dial Transplant 2011;26:3882–3888.

8 Davenport A, Tolwani A: Citrate anticoagulation for continuous renal replacement therapy (CRRT) in patients with acute kidney injury admitted to the intensive care unit. NDT Plus 2009;2:439–447.

9 Wu MY, Hsu YH, Bai CH, Lin YF, Wu CH, Tam KW: Regional citrate versus heparin anticoagulation for continuous renal replacement therapy: a meta-analysis of randomized controlled trials. Am J Kidney Dis 2012;59: 810–818.

10 Liu C, Mao Z, Kang H, Hu J, Zhou F: Regional citrate versus heparin anticoagulation for continuous renal replacement therapy in critically ill patients: a meta-analysis with trial sequential analysis of randomized controlled trials. Crit Care 2016;20:144.

11 Betjes MG, van Oosterom D, van Agteren M, van de Wetering J: Regional citrate versus heparin anticoagulation during venovenous hemofiltration in patients at low risk for bleeding: similar hemofilter survival but significantly less bleeding. J Nephrol 2007;20: 602–608.
12 Morabito S, Pistolesi V, Tritapepe L, Zeppilli L, Polistena F, Strampelli E, Pierucci A: Regional citrate anticoagulation in cardiac surgery patients at high risk of bleeding: a continuous veno-venous hemofiltration protocol with a low concentration citrate solution. Crit Care 2012;16:R111.
13 Oudemans-van Straaten HM, Bosman RJ, Koopmans M, van der Voort PH, Wester JP, van der Spoel JI, Dijksman LM, Zandstra DF: Citrate anticoagulation for continuous venovenous hemofiltration. Crit Care Med 2009; 37:545–552.
14 Hetzel GR, Schmitz M, Wissing H, Ries W, Schott G, Heering PJ, Isgro F, Kribben A, Himmele R, Grabensee B, Rump LC: Regional citrate versus systemic heparin for anticoagulation in critically ill patients on continuous venovenous haemofiltration: a prospective randomized multicentre trial. Nephrol Dial Transplant 2011;26:232–239.
15 Gattas DJ, Rajbhandari D, Bradford C, Buhr H, Lo S, Bellomo R: Randomized controlled trial of regional citrate versus regional heparin anticoagulation for continuous renal replacement therapy in critically ill adults. Crit Care Med 2015;43:1622–1629.
16 Monchi M, Berghmans D, Ledoux D, Canivet JL, Dubois B, Damas P: Citrate vs. heparin for anticoagulation in continuous venovenous hemofiltration: a prospective randomized study. Intensive Care Med 2004;30:260–265.
17 Kutsogiannis DJ, Gibney RT, Stollery D, Gao J: Regional citrate versus systemic heparin anticoagulation for continuous renal replacement in critically ill patients. Kidney Int 2005;67:2361–2367.
18 Schilder L, Nurmohamed SA, Bosch FH, Purmer IM, den Boer SS, Kleppe CG, Vervloet MG, Beishuizen A, Girbes AR, Ter Wee PM, Groeneveld AB; CASH Study Group: Citrate anticoagulation versus systemic heparinisation in continuous venovenous hemofiltration in critically ill patients with acute kidney injury: a multi-center randomized clinical trial. Crit Care 2014;18:472.
19 Stucker F, Ponte B, Tataw J, Martin PY, Wozniak H, Pugin J, Saudan P: Efficacy and safety of citrate-based anticoagulation compared to heparin in patients with acute kidney injury requiring continuous renal replacement therapy: a randomized controlled trial. Crit Care 2015;19:91.
20 Fealy N, Baldwin I, Bellomo R: The effect of circuit "down-time" on uraemic control during continuous veno-venous haemofiltration. Crit Care Resusc 2002;4:266–270.
21 Morgan TJ: The Stewart approach – one clinician's perspective. Clin Biochem Rev 2009; 30:41–54.
22 Jacobs R, Honore PM, Diltoer M, Spapen HD: Chloride content of solutions used for regional citrate anticoagulation might be responsible for blunting correction of metabolic acidosis during continuous veno-venous hemofiltration. BMC Nephrol 2016;17:119.
23 Khadzhynov D, Schelter C, Lieker I, Mika A, Staeck O, Neumayer HH, Peters H, Slowinski T: Incidence and outcome of metabolic disarrangements consistent with citrate accumulation in critically ill patients undergoing continuous venovenous hemodialysis with regional citrate anticoagulation. J Crit Care 2014;29:265–271.
24 Fiaccadori E, Pistolesi V, Mariano F, Mancini E, Canepari G, Inguaggiato P, Pozzato M, Morabito S: Regional citrate anticoagulation for renal replacement therapies in patients with acute kidney injury: a position statement of the work group "renal replacement therapies in critically ill patients" of the Italian society of nephrology. J Nephrol 2015;28:151–164.
25 Slowinski T, Morgera S, Joannidis M, Henneberg T, Stocker R, Helset E, Andersson K, Wehner M, Kozik-Jaromin J, Brett S, Hasslacher J, Stover JF, Peters H, Neumayer HH, Kindgen-Milles D: Safety and efficacy of regional citrate anticoagulation in continuous venovenous hemodialysis in the presence of liver failure: the liver citrate anticoagulation threshold (L-CAT) observational study. Crit Care 2015;19:349.
26 Schultheiß C, Saugel B, Phillip V, Thies P, Noe S, Mayr U, Haller B, Einwächter H, Schmid RM, Huber W: Continuous venovenous hemodialysis with regional citrate anticoagulation in patients with liver failure: a prospective observational study. Crit Care 2012;16:R162.

27 Kribben A, Gerken G, Haag S, Herget-Rosenthal S, Treichel U, Betz C, Sarrazin C, Hoste E, Van Vlierberghe H, Escorsell A, Hafer C, Schreiner O, Galle PR, Mancini E, Caraceni P, Karvellas CJ, Salmhofer H, Knotek M, Ginès P, Kozik-Jaromin J, Rifai K; HELIOS Study Group: Effects of fractionated plasma separation and adsorption on survival in patients with acute-on-chronic liver failure. Gastroenterology 2012;142:782–789.e3.

28 Bakker AJ, Boerma EC, Keidel H, Kingma P, van der Voort PH: Detection of citrate overdose in critically ill patients on citrate-anticoagulated venovenous haemofiltration: use of ionised and total/ionised calcium. Clin Chem Lab Med 2006;44:962–966.

29 Santoro A, Faenza S, Mancini E, Ferramosca E, Grammatico F, Zucchelli A, Facchini MG, Pinna AD: Prometheus system: a technological support in liver failure. Transplant Proc 2006;38:1078–1082.

30 Mancini E, Santoro A: Dialisis hepatica extracorporea en la unidad de cuitados criticos; in Villar SR, Marbad Libros SL (eds): Cuidados Criticos: Protocols, 2011, pp 259–271.

Davide Ricci, MD
U.O.C. Nefrologia, Dialisi e Ipertensione
Policlinico S. Orsola-Malpighi
Via Palagi 9, IT–40138 Bologna (Italy)
E-Mail davide.ricci@aosp.bo.it

La Manna G, Ronco C (eds): Current Perspectives in Kidney Diseases.
Contrib Nephrol. Basel, Karger, 2017, vol 190, pp 31–42 (DOI: 10.1159/000468909)

Coupled Plasma Filtration Adsorption Application for Liver and Thyroid Toxins

Gabriele Donati · Irene Capelli · Anna Laura Croci Chiocchini · Nicolò Natali · Anna Scrivo · Gaetano La Manna

Department of Experimental Diagnostic and Specialty Medicine (DIMES); Nephrology, Dialysis and Renal Transplant Unit, St. Orsola Hospital, University of Bologna, Bologna, Italy

Abstract

Coupled plasma filtration and adsorption (CPFA) is a detoxification system that combines a plasma adsorption circuit and a continuous renal replacement therapy circuit. Its main application is for sepsis and septic shock with or without acute renal failure. Several recent studies have suggested that CPFA can reduce the mortality when the volume of plasma absorbed on the styrenic resin is at least >0.18 L/kg/day. At present, new applications for CPFA are under investigation, also in patients without significant kidney failure. We report here a successful case of CPFA use during acute liver failure, with a complete recovery of liver function in a patient after severe cholangitis and relapsing hemolytic anemia. The resin enabled the removal of bilirubin and protein-bound toxins, while the hemofilter removed the hydrophilic toxins such as ammonia and non-protein–bound toxins as free bilirubin. We also describe a second case of CPFA application during thyrotoxicosis to achieve free triiodothyronine (FT3) and free thyroxin (FT4) adsorptions. The CPFA efficacy seems to exceed that obtained by plasma exchange (PEX) as to FT3 and FT4 adsorptions. The resin allowed the adsorption of FT3 and FT4. The role of the hemofilter is to enhance the hemodynamic tolerance of the extracorporeal treatment and remove water-soluble toxins. The reduced duration of CPFA treatments, in case of normal renal function, is confirmed by the assessment of the resin cartridge saturation. Thus, multipurpose CPFA can play a role in the case of resistance to current medical therapy or as a bridge to liver transplantation or thyroidectomy.

Introduction

Coupled plasma filtration adsorption (CPFA) is a detoxification system that combines a plasma adsorption circuit, by means of a cartridge containing styrenic resin, with a continuous renal replacement therapy (CRRT) circuit. This technique is available on the Bellco Amplya CRRT machine. The first part of the system consisted of a Micropes® plasmafilter with a surface area of 0.45 m^2, a membrane cut-off value of 800,000. The sieving coefficient for albumin is 1.0. The percentage of plasma extraction from the whole blood ranges between 10 and 20%. The plasma is extracted from the plasma filter to pass through 140 mL of divinylbenzene styrenic macroporous resin (Mediasorb®) with a surface area of 700 m^2/g and an estimated adsorbing area of 50,000 m^2. The diameter of the resin particles ranges between 50 and 100 microns; while the porous diameter in the resin particles is about 30 nm. After adsorption, the purified plasma is returned to the whole blood that, in turn, passes through a hemofilter. The hemofilter is a polyphenylene high flux filter (HFT14, Kuf 41 mL/h/mm Hg, surface area 1.4 m^2, sieving coefficient for albumin 0.01) where weight loss and convective depuration take place. An infusion rate of up to 2,500 mL/h is allowed in the post-dilution method. The priming volume of the whole CPFA circuit is approximately 300 mL. CPFA allows an improvement of blood pressure and inflammatory response during sepsis and multiorgan failure [1]. Many authors observed that these clinical effects depend on the improvement of the ratio between anti-inflammatory and pro-inflammatory cytokines [2–4]. Ronco et al. [5] in 2002 compared the CPFA treatment to continuous veno-venous hemodiafiltration in 10 patients affected by septic shock. An increase of arterial blood pressure and a significant reduction of norepinephrine dosage were observed with CPFA. Recently, Livigni et al. [6] enrolled 330 adult patients with septic shock; 192 were randomized to either having CPFA added to the standard care, or not. CPFA was to be performed daily for 5 days, lasting at least 10 h/day. The trial was stopped early as nearly 50% of the patients could not complete the treatment regimen due to technical reasons (primarily coagulation). The trial did observe that patients with the highest amounts of treated plasma appeared to have a survival benefit. The technique underwent further modifications to improve the technical aspects and anticoagulation, and a new trial is now underway with a high dose of plasmas >0.18 L/kg/day.

At present, new applications for CPFA are under investigation also in patients without significant kidney failure or sepsis.

Case Reports

Case 1

A 51-year-old man with systemic lupus erythematosus, positive LAC test, Fisher-Evans syndrome (autoimmune hemolytic anemia and thrombocytopenia), and previous cholecystectomy was admitted for jaundice, abdomen pain, and fever (38 °C). His blood pressure and heart rate were 120/60 mm Hg and 105 beats/min, respectively. The abdomen was protruding but treatable. The Murphy's sign was positive. Peristalsis was present and valid. White blood cells were 7,100/μL, hemoglobin was 7.4 g/dL, platelets 33,000/μL, INR 1.51, creatinine 1.06 mg/dL, total bilirubin 27.17 mg/dL, direct bilirubin 21.65 mg/dL, indirect bilirubin 5.52 mg/dL, AST 323 U/L, ALT 474 U/L, and LDH 342 U/L. The CT scan of the abdomen with iodinate contrast showed an enlargement of the intrahepatic bile ducts and of the choledochus due to 7 radiopaque stones. During the next 24 h, platelets decreased to 17,000/μL, total bilirubin increased to 53.26 mg/dL, direct bilirubin was 41.17, indirect bilirubin was 11.09 mg/dL, C-reactive protein was 10.52 mg/dL. The patient was transfused with 600 mL of red blood cells and 600 mL of platelets. The consultant hematologist suggested methylprednisolone 1 mg/kg/die due to the relapse of the Fisher-Evans syndrome. An endoscopic retrograde cholangiopancreatography with papillotomy of the Water's ampulla was carried out. The drainage of bile stones, bile sludge, and pus was obtained. On the third day, the total bilirubin increased to 67.34 mg/dL (direct bilirubin 54.68 mg/dL, indirect bilirubin 12.66 mg/dL), hemoglobin was 9 g/dL, platelets 13,000/μL, creatinine 1.27 mg/dL. A new CT scan of the abdomen ruled out any enlargement of the bile ducts. During the fourth day, a nephrology consultant was requested for consideration of an extracorporeal detoxification treatment. The patient was referred to the Dialysis Unit and underwent 2 CPFA treatments lasting 6 h each. Unfractioned heparin was used as an anticoagulant of the extracorporeal circuit. After the first treatment, 8 L of plasma were treated with resin adsorption. Blood pressure and heart rate were 110/60 mm Hg and 100 beats/min, respectively. Total bilirubin decreased from 71.20 (direct 54.58 mg/dL, indirect 23.27 mg/dL) to 41.54 mg/dL (direct 29.86 mg/dL, indirect 11.65 mg/dL), the total bilirubin reduction ratio was 41%, ammonia decreased from 67.4 to 21.1 μmol/L, the ammonia reduction ratio was 68%, hemoglobin was 8.4 g/dL, platelets 18,000/μL, and creatinine 1.22 mg/dL. Figure 1 shows the total bilirubin values checked before and after the resin cartridge during the first CPFA treatment. Table 1 shows the total bilirubin blood levels at the checking time during CPFA and bilirubin plasma values before and after the resin cartridge: the bilirubin blood levels are very similar to bilirubin plasma levels pre-cartridge. On the fifth day, the patient underwent a second CPFA treatment. A plasma volume of 8 L was treated with resin adsorption. Blood pressure and heart rate were 120/70 mm Hg and 90 beats/min, respectively. Total bilirubin decreased from 20.27 (direct 15.41 mg/dL, indirect 4.86 mg/dL) to 12.23 mg/dL (direct 10.53 mg/dL, indirect 1.70 mg/dL), the total bilirubin reduction ratio was 39%, ammonia decreased from 37.3 to 11.2 μmol/L, the ammonia reduction ratio was 69%, hemoglobin was 7.5 g/dL, platelets

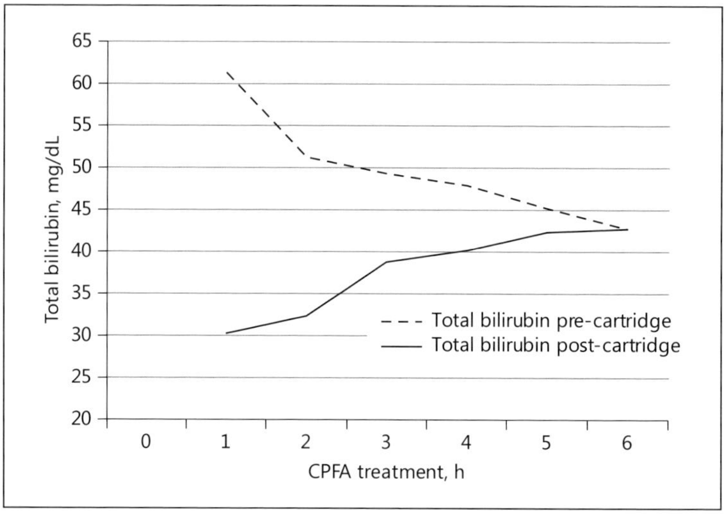

Fig. 1. Pre- and post-cartridge total bilirubin values during the first CPFA treatment.

Table 1. Total bilirubin and FT3 levels in blood and plasma at the checking times during the first CPFA treatment

CPFA, h	Total bilirubin blood levels, mg/dL	Total bilirubin plasma levels pre-cartridge, mg/dL	Total bilirubin plasma levels post-cartridge, mg/dL
0	71.2		
1	61.2	61.2	30.2
2	51.9	51.2	32.3
3	49.5	49.3	38.7
4	48.1	47.9	40.1
5	45.2	45.1	42.3
6	42.7	42.7	42.7
CPFA, h	**FT3 blood levels, pg/mL**	**FT3 plasma levels pre-cartridge, pg/mL**	**FT3 plasma levels post-cartridge, pg/mL**
0	11.1		
1	9.5	6.5	1.9
2	9.0	6.4	1.8
3	7.8	5.9	2.2
4	7.6	5.7	2.5
5	7.4	5.7	2.7
6	7.2	5.7	2.7

30,000/μL, and creatinine 1.22 mg/dL. After 12 days, the patient was discharged, his total bilirubin was 3.1 mg/dL, hemoglobin 8.5 g/dL, platelets 40,000/μL, creatinine 1.06 mg/dL.

Case 2

A 61-year-old man with diabetes and ischemic cardiomyopathy, and with an ejection fraction of 35% and atrial fibrillation, ceased the amiodarone therapy 4 months prior to skin reaction. Two months after amiodarone discontinuation, he started thiamazole 30 mg/die for the diagnosis of thyrotoxicosis. One month later, he was re-admitted to the Cardiology Unit for relapsing NSTEMI (non-ST elevation myocardial infarction) and atrial fibrillation with high ventricular response. Blood pressure and heart rate were 90/50 mm Hg and 150 beats/min, respectively. A coronary angiography was carried out; angioplasty was not considered despite 90% ostial stenosis of the circumflex branch and a 40% stenosis of the anterior interventricular branch. Continuous i.v. metoprolol was administered. Due to the worsening of thyrotoxicosis, the patient started the following therapy: methylprednisolone 1 mg/kg/die, thiamazole 40 mg/die, sodium perchlorate 800 mg/die, cholestyramine 12 mg/die. The thyrotoxicosis was resistant to medical therapy: the thyroid-stimulating hormone levels were <0.01 IU/L, free triiodothyronine (FT3) was 11.3 pg/mL, free thyroxin (FT4) was 58 pg/mL. Two further episodes of NSTEMI were diagnosed after the admission. The risk of death with thyroidectomy exceeded 50%. Creatininemia was 0.5 mg/dL. A nephrology consult was requested for consideration of an extracorporeal detoxification treatment. The patient was referred to the Dialysis Unit and underwent 2 consecutive CPFA treatments lasting 6 h each. After the first treatment, FT3 decreased from 11.1 to 7.2 pg/mL, FT4 from >60 to 56.9 pg/mL, the FT3 reduction ratio was 36%, the FT4 reduction ratio was not calculated because the initial FT4 value exceeded the upper sensitivity limit of the laboratory assay. Figures 2 and 3 show, respectively, FT3 and FT4 values checked before and after the resin cartridge. Table 1 shows the FT3 blood levels at the checking time during CPFA and FT3 plasma values before and after the resin cartridge: the FT3 blood levels are different compared to FT3 plasma levels pre-cartridge. Hemoglobin was 11.2 g/dL and platelets were 138,000/μL. Eight liters of plasma were treated with resin adsorption. Blood pressure and heart rate were 100/70 mm Hg and 150 beats/min, respectively. They remained stable during CPFA. During the second CPFA treatment, FT3 decreased from 9.9 to 6.9 pg/mL, FT4 from 59.3 to 57.4 pg/mL, the FT3 reduction ratio was 30%, the FT4 reduction ratio was 3%, and hemoglobin was 11.4 g/dL. Amiodarone dosage was carried out by means of the HPLC method. The amiodarone value at the commencement of CPFA treatment was 0.14 μg/mL (the therapeutic range is between 0.4 and 1.0 μg/mL).

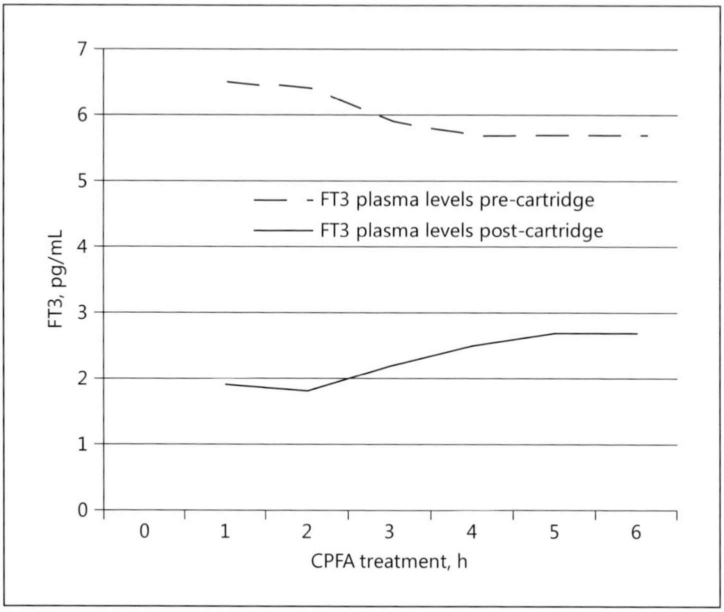

Fig. 2. Pre- and post-cartridge FT3 values during the first CPFA treatment.

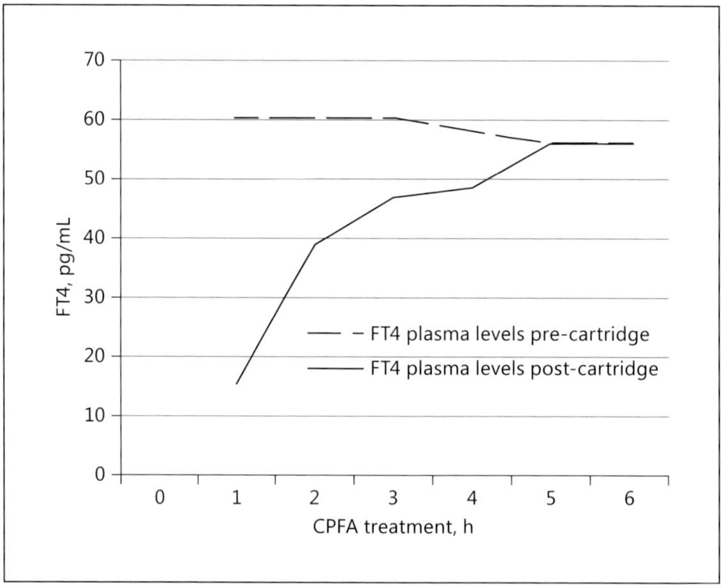

Fig. 3. Pre- and post-cartridge FT4 values during the first CPFA treatment.

At the end of CPFA treatment, it was 0.09 μg/mL. The amiodarone reduction rate was 35%. Blood pressure and heart rate were 90/50 mm Hg and 140 beats/min, respectively, and they remained stable during CPFA. Eight liters of plasma were treated with resin adsorption. Then, 5 plasma exchange (PEX) were carried out. Before the first PEX, FT3 and FT4 were 7.4 and 56.4 pg/mL, respectively. After the last PEX, FT3 and FT4 were 6.2 and 52.6 pg/mL, respectively. Hemoglobin was 7.8 g/dL, platelets were 86,000/μL. Two days after the last PEX, the patient died.

Discussion

New CPFA Application: Liver Failure

The efficacy of the hepatic toxin removal depends on the ability of the detoxification system to extract albumin-bound toxins by means of adsorption. The adsorption comprises the consolidation on the surface of a solid material of molecules floating in a fluid after matching the fluid with the solid. The consolidation is due to weak bonds, as in the case of Van der Waals and hydrophobic bonds. The selectivity of adsorption occurs when the albumin molecules easily pass through the sorbent while the sorbent draws the albumin carrying toxins into the pores of the sorbent [7]. The hepatic toxins often are hydrophobic, with a molecular weight <1,000 daltons (bilirubin = 406 daltons, cholic acid = 283 daltons, chenodeoxycholic acid = 272 daltons). The albumin structure consists of 3 domains, which allows a multi-binding protein transport [8]. Regarding bilirubin, its protein binding is a physiological protective mechanism to prevent the free bilirubin diffusion and toxicity in human tissues. The amount of bilirubin production is about 300 g/day. The molecule most tightly bound to albumin is non-conjugated bilirubin: the amount of non-conjugated bilirubin not bound to albumin is <0.1% [8]. Non-conjugated bilirubin is currently considered as “indirect” bilirubin because it gives an indirect reaction with standard diazo reagents [9]. Conjugated bilirubin is also bound to albumin, but with a reduced affinity. The albumin-bilirubin bond occurs by means of covalent or non-covalent strengths. The 2 bilirubin binding sites on albumin show different constant of association (Ka). The primary site has a high affinity, its Ka is 55–68 μmol/L and it can bind only 1 mole of bilirubin. The secondary site has a low affinity, its Ka is 4.4–5.0 μmol/L and it can bind more than 1 mole of bilirubin [10]. A gap between 20 and 50 ångström between albumin and the sorbent is needed to move bilirubin from albumin to the sorbent [8, 10].

The CPFA efficacy on bilirubin adsorption is based on hydrophobic polystyrene-divinylbenzene copolymer matrices that have a mean pore size of 30 nm.

Harm et al. [11] reported that the hydrophobic polystyrene-divinylbenzene copolymer of CPFA resin cartridge (type B copolymer: mean diameter 128 micron, pore size 30 nm, and surface area of 700 m^2/g of resin) binds higher amount of bilirubin than the amount of bilirubin bound with other kinds of polystyrene-divinylbenzene copolymer matrices tested (type A copolymer: mean diameter 120 micron, pore size 15 nm, and surface area of 900 m^2/g of resin; type C copolymer: mean diameter 37 micron, pore size 100 nm and surface area of 200 m^2/g of resin).

Few authors in vivo assessed the bilirubin adsorbing ability of the neutral styrenic resin of CPFA. Caroleo et al. [12] reported the first case of hepatic toxin removal by means of CPFA in a patient affected by acute liver failure due to cardiogenic shock. Maggi et al. [13] reported 2 further cases where bilirubin removal was obtained in 2 patients with delayed graft function after liver transplantation. The patients achieved a 40% bilirubin reduction after 3 treatments with CPFA lasting 6 h each.

Here we evaluated the CPFA efficacy in a patient affected by acute liver failure due to severe post obstructive cholangitis and relapsing hemolytic anemia. The resin has a pivotal role in removing bilirubin and protein-bound toxins, but it also reduces the inflammatory mediators due to cholangitis. The role of the hemofilter was the removal of hydrophilic toxins such as ammonia and non-protein bound toxins as free bilirubin. The bilirubin and ammonia reduction ratios per session were considered in accordance with our previous paper [14]. The duration of CPFA treatments was confirmed by the assessment of the resin cartridge saturation: after 6 h of treatment, the bilirubin values after the resin cartridge were similar to those observed before the resin.

New CPFA Application Thyrotoxicosis

The thyrotoxicosis is a medical emergency caused by an exacerbation of the hyperthyroid state characterized by the decompensation of one or more organ systems. Timely recognition and aggressive treatment are essential in limiting the morbidity and mortality associated with the disease. Commonly, the crisis has an abrupt onset, and is triggered by a precipitating factor, namely acute trauma, infectious diseases, ketoacidosis, thyroidal surgery, 131-I radio-metabolic treatment, administration of iodine-containing materials, parturition. The principles of the therapeutic management of thyrotoxic crisis are: the reduction of circulating thyroid hormone levels; the inhibition of the peripheral effects of circulating thyroid hormones; the supportive care to reverse the systemic decompensation, and the treatment of the underlying precipitating event. The thyroid storm is the extreme manifestation of thyrotoxicosis [15, 16].

When clinical deterioration occurs during a thyroid storm, despite the use of the medical therapy, a therapeutic option to be considered is the removal of the thyroid hormone from circulation. Plasmapheresis, charcoal hemoperfusion, resin hemoperfusion, and PEX have been found to be effective in rapidly reducing the thyroid hormone levels in thyroid storm [16]. In healthy subjects, approximately 0.03% of the total serum T4 (molecular weight 776 daltons) and 0.3% of the total serum T3 (molecular weight 651 daltons) are present in free or unbound form. The total serum T3 production rate is 50 nmol/day, the relative metabolic potency is 1.0, the distribution volume is 40 L, and the half-life is 18 h. The total serum T4 production rate is 110 nmol/day, the relative metabolic potency is 0.3, the distribution volume is 10 L, and the half-life is 6.7 days [17]. The major serum thyroid hormone-binding proteins are thyroxine-binding globulin (TBG), transthyretin (TTR), and albumin. The molecular weights of TBG, TTR, and albumin are 54,000, 55,000, and 69,000 daltons, respectively. In terms of their relative abundance in serum, albumin levels are approximately 100-fold the molar concentration of TTR and 2,000-fold that of TBG [18]. However, from the viewpoint of the association constants for T4, TBG has the highest affinity, which is 50-fold higher than that of TTR and 7,000-fold higher than that of albumin. As a result, TBG binds 75% of serum T4, while TTR and albumin bind only 20 and 5%, respectively [17]. PEX has been described as an effective method of last therapeutic resort for therapy-resistant thyrotoxicosis [19] and thyroxin intoxication [20]. In both situations, substantially higher FT4 levels have been described based on an enhanced endogenous production or intake. Due to the ongoing endogenous production, PEX is rather considered a temporal relief therapy before definitive surgery. Given the current observation of just small reductions in FT4 and FT3 and only insignificant reductions of FT4 in one retrospective study following PEX, the utility of such a procedure has to be questioned, although it may be a life-saving procedure in the case of a thyroid storm [19, 21].

An alternative technique to PEX is resin adsorption, in particular it is advisable due to the high hemodynamic tolerance of this kind of extracorporeal treatment [22]. Burman et al. [23] assessed the efficacy of a resin hemoperfusion system in decreasing T3, T4, and FT4. The study was carried out on an experimental model of thyrotoxic dogs that were submitted to resin hemoperfusion treatments of 2-h duration. The sorbent used was a polymeric neutral resin (mean diameter 300–450 micron, mean surface area 750 m^2/g; mean pore size 5 nm). The mean serum T3, T4, and FT4 decreased by 39, 35, and 46%, respectively.

The present case of CPFA use considers a patient affected by thyrotoxicosis due to amiodarone intoxication diagnosed before the present admission. The intoxication might be enhanced by the iodinate medium contrast administered during the coronary angiography and was resistant to maximized medical therapy. In the early phase of the extracorporeal treatment, CPFA was preferred to foster the hemodynamic tolerance to the extracorporeal treatment: during CPFA, no leakage of albumin or plasma takes place, while hemofiltration stabilizes the plasma osmolarity and decreases the body core temperature resulting in a better preservation of blood pressure [24–26]. During the 2 CPFA treatments, a 30% reduction ratio in FT3 values occurred but the resin saturation by FT3 was not achieved. This finding suggests a CPFA treatment lasting more than 6 h to obtain resin saturation by FT3. Nonetheless the rate of T3 shift from the protein-bound form to the free form seems to be lower than the rate of FT3 removal by the resin. In this way, we can explain the reduction of blood FT3. As to FT4 values, the resin showed a high affinity with a rapid saturation: the FT4 levels post-resin cartridge already achieved the FT4 levels after 4 h of treatment. Such behavior did not cause a significant reduction in the circulating FT4 values, indicating a rapid shift from the bound T4 pool to the free T4 fraction.

Amiodarone removal during CPFA was assessed. The withdrawal of the drug was carried out 4 months before the present hospital admission. Nonetheless, amiodarone was still detected in blood and removed by CPFA. Plasma amiodarone concentrations decrease rapidly (initial half-life: 2.5–10 days) during the first 12–21 days after discontinuation of long-term therapy. However, the large volume of distribution (5,000 L) and the prolonged terminal elimination half-life (about 40 days) of the drug can be related to its presence in the blood stream [27].

Conclusions

CPFA can play a role in the case of endogenous toxin accumulation when the resistance to the current medical therapy occurs. The main advantage of this approach is its ability to treat a higher amount of plasma in comparison to PEX without plasma or albumin replacement and with a high hemodynamic tolerance. This advantage can be useful, especially for those toxins with a high volume of distribution that can benefit from a prolonged extracorporeal detoxification. The essential condition for the efficiency of CPFA is its timely application. The depurative efficiency of the tested resin should be improved and tailored to the different kinds of toxins to be removed. These case reports should encourage new studies with more patients to confirm our results on a large number of protein-bound toxins.

References

1 Ronco C, Brendolan A, d'Intini V, Ricci Z, Wratten ML, Bellomo R: Coupled plasma filtration adsorption: rationale, technical development and early clinical experience. Blood Purif 2003;21:409–416.
2 Tetta C, Cavaillon JM, Schulze M, Ronco C, Ghezzi PM, Camussi G, Serra AM, Curti F, Lonnemann G: Removal of cytokines and activated complement components in an experimental model of continuous plasma filtration coupled with sorbent adsorption. Nephrol Dial Transplant 1998;13:1458–1464.
3 Mao HJ, Yu S, Yu XB, Zhang B, Zhang L, XU XR, Wang XY, Xing CY: Effects of coupled plasma filtration adsorption on immune function of patients with multiple organ dysfunction syndrome. Int J Artif Organs 2009; 32:31–38.
4 Cole L, Bellomo R, Davenport P, Tipping P, Uchino S, Tetta C, Ronco C: The effect of coupled haemofiltration and adsorption on inflammatory cytokines in an ex vivo model. Nephrol Dial Transplant 2002;17:1950–1956.
5 Ronco C, Brendolan A, Lonnemann G, Bellomo R, Piccinni P, Digito A, Dan M, Irone M, La Greca G, Inguaggiato P, Maggiore U, De Nitti C, Wratten ML, Ricci Z, Tetta C: A pilot study of coupled plasma filtration with adsorption in septic shock. Crit Care Med 2002;30:1250–1255.
6 Livigni S, Bertolini G, Rossi C, Ferrari F, Giardino M, Pozzato M, Remuzzi G: Efficacy of coupled plasma filtration adsorption (CPFA) in patients with septic shock: a multicenter randomised controlled clinical trial. BMJ Open 2014;4:e003536.
7 Amandola G, Terreni V: Cromatografia; in Amandola G, Terreni V (eds): Analisi Chimica Strumentale e Tecnica, ed VI. Milano, Masson Scuola, 1995, pp 366–370.
8 Hughes RD: Review of methods to remove protein-bound substances in liver failure. Int J Artif Organs 2002;25:911–917.
9 Weiss JS, Gautam A, Lauff JJ, Sundberg MW, Jatlow P, Boyer JL, Seligson D: The clinical importance of a protein-bound fraction of serum bilirubin in patients with hyperbilirubinemia. N Eng J Med 1983;309:147–150.
10 Gong D, Cruz D, Ronco C: Depurative capacity of molecular adsorbent recycling system (MARS): a focus on bilirubin removal. Int J Artif Organs 2008;31:875–881.
11 Harm S, Falkenhagen D, Hartmann J: Pore size – a key property for selective toxin removal in blood purification. Int J Artif Organs 2014;37:668–678.
12 Caroleo S, Rubino AS, Tropea F, Bruno O, Vuoto D, Amantea B, Renzulli A: Coupled plasma filtration adsorption reduces serum bilirubine in a case of acute hypoxic hepatitis secondary to cardiogenic shock. Int J Artif Organs 2010;33:749–752.
13 Maggi U, Nita G, Gatti S, Antonelli B, Paolo R, Como G, Messa P, Rossi G: Hyperbilirubinemia after liver transplantation: the role of coupled plasma filtration adsorption. Transplant Proc 2013;45:2715–2717.
14 Donati G, La Manna G, Cianciolo G, Grandinetti V, Carretta E, Cappuccilli M, Panicali L, Iorio M, Piscaglia F, Bolondi L, Colì L, Stefoni S: Extracorporeal detoxification for hepatic failure using molecular adsorbent recirculating system: depurative efficiency and clinical results in a long-term follow-up. Artif Organs 2014;38:125–134.
15 Migneco A, Ojetti V, Testa A, De Lorenzo A, Gentiloni Silveri N: Management of thyrotoxic crisis. Eur Rev Med Pharmacol Sci 2005; 9:69–74.
16 Nayak B, Burman K: Thyrotoxicosis and thyroid storm. Endocrinol Metab Clin North Am 2006;35:663–686, vii.
17 Salvatore D, Davies TF, Schlumberger ML, Hay ID, Larsen PR: Evaluation of patients with thyroid disorders; in Melmed S, Polonsky KS, Larsen PR (eds): Williams Textbook of Endocrinology. Philadelphia, Elsevier, 2015, pp 345–346.
18 Oppenheimer JH: Role of plasma proteins in the binding, distribution and metabolism of the thyroid hormones. N Engl J Med 1968; 278:1153–1162.
19 Ezer A, Caliskan K, Parlakgumus A, Belli S, Kozanoglu I, Yildirim S: Preoperative therapeutic plasma exchange in patients with thyrotoxicosis. J Clin Apher 2009;24:111–114.
20 Binimelis J, Bassas L, Marruecos L, Rodriguez J, Domingo ML, Madoz P, Armengol S, Mangues MA, de Leiva A: Massive thyroxine intoxication: evaluation of plasma extraction. Intensive Care Med 1987;13:33–38.

21 Jha S, Waghdhare S, Reddi R, Bhattacharya P: Thyroid storm due to inappropriate administration of a compounded thyroid hormone preparation successfully treated with plasmapheresis. Thyroid 2012;22:1283–1286.
22 Guidet B, Luquel L, De Labriolle Vaylet C, Offenstadt G: Hemodynamic tolerance and plasma volume variations during plasma exchange. Replacement fluid: albumin alone or albumin plus gelatin? Int J Artif Organs 1987; 10:325–330.
23 Burman KD, Yeager HC, Briggs WA, Earll JM, Wartofsky L: Resin hemoperfusion: a method of removing circulating thyroid hormones. J Clin Endocrinol Metab 1976;42:70–78.
24 Van Biesen W, Lameire N: Slow efficiency daily dialysis and hybrid renal replacement therapies for acute renal failure in the intensive care unit; in Murray PT, Brady HR, Hall JB (eds): Intensive Care Nephrology. London, Taylor & Francis, 2007, pp 308–323.
25 Colì L, La Manna G, Comai G, Ursino M, Ricci D, Piccari M, Locatelli F, Di Filippo S, Cristinelli L, Bacchi M, Balducci A, Aucella F, Panichi V, Ferrandello FP, Tarchini R, Lambertini D, Mura C, Marinangeli G, Di Loreto E, Quarello F, Forneris G, Tancredi M, Morosetti M, Palombo G, Di Luca M, Martello M, Emiliani G, Bellazzi R, Stefoni S: Automatic adaptive system dialysis for hemodialysis-associated hypotension and intolerance: a noncontrolled multicenter trial. Am J Kidney Dis 2011;58:93–100.
26 Colì L, Ursino M, Donati G, Cianciolo G, Soverini ML, Baraldi O, La Manna G, Feliciangeli G, Scolari MP, Stefoni S: Clinical application of sodium profiling in the treatment of intradialytic hypotension. Int J Artif Organs 2003;26:715–722.
27 Uzzan B, Pussard E, Leon A, Bekhechi D, Krivitzky A, Modigliani E, Perret G, Vassy R, Berdeaux A, Giudicelli JF: The effects of plasmapheresis on thyroid hormone and plasma drug concentrations in amiodarone-induced thyrotoxicosis. Br J Clin Pharmacol 1991;31: 371–372.

Gaetano La Manna, MD, PhD
Department of Experimental, Diagnostic and Specialty Medicine (DIMES)
Nephrology, Dialysis and Transplantation Unit
St. Orsola Hospital, University of Bologna
Via G. Massarenti 9 (Pad. 15), IT–40138 Bologna (Italy)
E-Mail gaetano.lamanna@unibo.it

La Manna G, Ronco C (eds): Current Perspectives in Kidney Diseases.
Contrib Nephrol. Basel, Karger, 2017, vol 190, pp 43–57 (DOI: 10.1159/000468911)

Extracorporeal Sorbent Technologies: Basic Concepts and Clinical Application

William R. Clark[a] • Fiorenza Ferrari[b] • Gaetano La Manna[c] • Claudio Ronco[b, d]

[a]School of Chemical Engineering, Purdue University, West Lafayette, IN, USA; [b]International Renal Research Institute of Vicenza (IRRIV), San Bortolo Hospital, Vicenza, [c]Department of Experimental Diagnostic and Specialty Medicine (DIMES); Nephrology, Dialysis and Renal Transplant Unit, St. Orsola Hospital, University of Bologna, Bologna, and [d]Department of Nephrology, San Bortolo Hospital, Vicenza, Italy

Abstract

Limitations imposed by the characteristics of some solutes and the structure of dialysis membranes have spurred new interest in the use of mechanisms beyond diffusion and convection for extracorporeal solute removal. Sorbents have been utilized for more than 50 years in extracorporeal blood treatments for specific purposes, and better understanding of their basic aspects may further expand the potential for their clinical application. In this chapter, the basic principles applying to sorbents are discussed, including composition and structure, along with the fundamental mechanisms of solute removal. The critical importance of sorbent biocompatibility is also highlighted. With these basic principles in mind, the clinical application of sorbents is discussed, with an emphasis on the use of hemoperfusion and coupled plasma filtration-adsorption for sepsis-related disorders. Finally, new sorbent-based clinical approaches for acute conditions and end-stage renal disease are presented, emphasizing that sorbent technologies may assume a larger role for a variety of clinical disorders in the future.

Introduction

Solute removal in hemodialysis and other blood purification techniques is mainly achieved by diffusion and convection. However, the limitations imposed by the characteristics of some solutes and the structure of dialysis membranes have spurred new interest in the use of further mechanisms of solute removal such as

Table 1. Development of sorbents in extracorporeal blood therapies

1850	First inorganic aluminosilicate (zeolites) used to exchange NH_4 and Ca
1910	Water softeners using zeolites display instability in the presence of mineral acids
1935	Adams and Holmes synthesize the first organic polymer ion exchange resin
1950	Application of synthetic porous polymers (styrene or acrylic acid based) (spherical beads: trade names of Amberlite, Duolite, Dowex, Ionac, and Purolite)
1960	Manipulation of physiochemical characteristics (commercial use)
1970	Application in blood purification techniques such as hemoperfusion
1980–2000	Improved design and coating for better hemocompatibility of adsorbent materials
2000 and beyond	Search for new sorbent materials and new possibilities of application

adsorption [1, 2]. Materials with high adsorptive capacity (sorbents) have been utilized for more than 50 years in extracorporeal blood treatments for specific purposes. The evolution in knowledge and clinical use of sorbents has been significant over the years and can be summarized in Table 1.

The analysis of the molecular structure of sorbents, as well as the study of the physiochemical mechanisms involved in the process of adsorption, are fascinating. A better understanding of these basic aspects may further expand the potential for the clinical application of sorbent materials [3, 4].

Basic Principles

In clinical settings, blood purification techniques achieve molecular separation primarily by membrane-based and adsorptive processes [5, 6]. When the former techniques using diffusion and convection are inadequate to remove the target molecules from the patient's blood, the use of sorbents and hemoperfusion (HP) may become an additional option for blood purification. In HP, blood is circulated through a unit (cartridge) containing the solid sorbent material. Solute removal and blood purification are obtained by absorption (binding) of molecules onto the sorbent particles.

Sorbents can be composed of synthetic or natural materials. In the past, the application of HP was limited by the relative bio-incompatibility of the sorbent material and the significant side effects derived from its contact with blood. HP sessions were often accompanied by chills, fever, cutaneous rush, thrombocytopenia, leukopenia, and aluminum leaching. Today, these reactions have become rare and can be prevented in 2 ways:

- In some techniques, plasma is separated from cells before being circulated through the sorbent bed. After the sorbent cartridge, blood is reconstituted so that red cells, white cells, and platelets never come in contact with the sorbent surface, and bio-incompatibility reactions are avoided.
- The sorbent material is made bio- or hemo-compatible by a specific coating process that covers the particles with bio-layers that are well tolerated by blood cells [7].

There is little debate that the use of sorbents is justified in poisoning or acute intoxications, for which HP is the treatment of choice in many instances due to the high affinity of the sorbent for the specific toxic molecule. (This aspect of HP is not the focus of this chapter.) However, the use of sorbents in chronic or acute blood purification techniques is still a matter of discussion. In particular, the additional value offered by adsorption must be counterbalanced by the increase in costs that are involved when sorbents are utilized. Nevertheless, the relative selectivity of adsorptive processes and the possibility of placing the sorbent in direct contact with blood may be seen as a further step toward increasing the efficiency and specificity of the blood purification process for certain types of solutes [8]. In particular, specific molecules can be targeted for removal by selective adsorption mechanisms. Furthermore, solutes with molecular size larger than the pore dimensions of membranes can be removed by direct adsorption onto the surface of the sorbent particles.

Sorbent Materials and Structure

To deliver an adequate adsorbent-based therapy, some important requirements must be fulfilled: (1) an effective, biocompatible, and safe sorbent material; (2) a sorbent cartridge with adequate design and structure; and (3) operating conditions allowing for optimal utilization of the available surface of the sorbent [9].

Sorbents are present in nature as raw materials or they can be synthetically produced in the laboratory. Natural sorbents such as zeolites (aluminum silicates) are inorganic polymers with remarkable porosity, deriving from their crystal structure, and can be synthetically modified to control the structure of

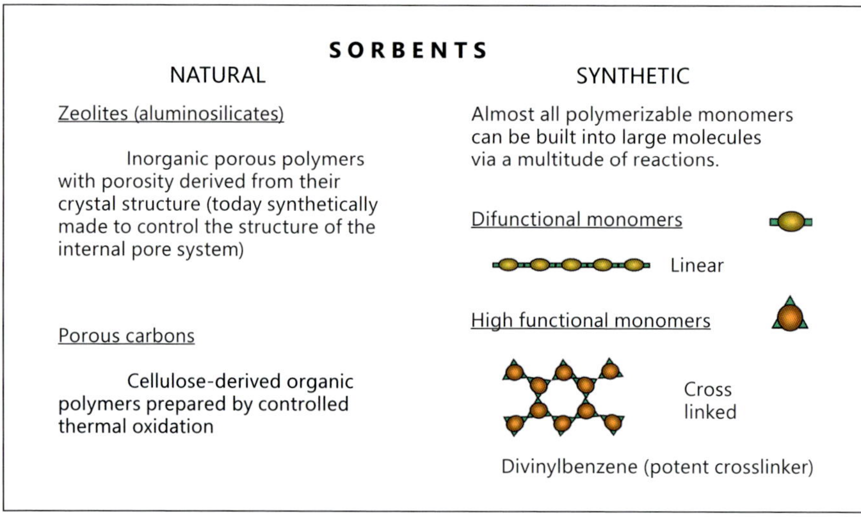

Fig. 1. Description of sorbent characteristics and distinction between natural and synthetic sorbents.

the internal pore system. Other typical sorbents such as porous carbons are cellulose-derived organic polymers prepared by controlled thermal oxidation (Fig. 1).

Different polymers of synthetic origin constitute the other class of sorbents. Almost all monomers susceptible to cross-linking can be transformed into large polymeric molecules via a multitude of reactions. Bifunctional monomers tend to aggregate in linear polymeric structures while highly functional monomers tend to polymerize in cross-linked structures. Divinylbenzene is a potent cross-linker frequently utilized to build polymeric sorbent molecules. Sorbent polymers can also be functionalized with chemical compounds to target specific molecules for adsorption.

Sorbents exist in granules, spheres, fibers, cylindrical pellets, flakes, and powder. They are solid particles with single particle diameters generally ranging between 50 μm and 1.2 cm. The surface area to volume ratio (S/V) is extremely high in sorbent particles, with an effective surface area varying from 300 to 1,200 m^2/g. They are also classified according to the size of the pores of the inner structure: (a) macroporous: pore size >500 Å (50 nm); (b) mesoporous: pore size 20–500 Å; and (c) microporous: pore size <20 Å.

The surface to volume ratio (S/V) is generally described by the following equation:

$$S/V = \pi d_p L\ (\pi d_p^2 L/4) = 4d_p \qquad (1)$$

where d_p is the pore diameter and L is the pore length. Considering fractional particle porosity (ε_p) and particle density (ρ_p), the specific surface area per unit of mass (S_g) is:

$$S_g = 4\ \varepsilon_p/\rho_p\ d_p \quad (2)$$

As an example of a clinically realistic application, if $\varepsilon_p = 0.5$, $\rho_p = 1$ g/cm^3 (1×10^6 g/m^3), and $d_p = 20$ Å (20×10^{-10} m), $S_g = 1{,}000$ m^2/g. In other words, 1 g of sorbent material provides a potential surface for adsorption of 1,000 m^2. Frequently, however, the available surface is not fully utilized since many factors contribute to limit the fraction of surface actually available for adsorption.

Requirements for a Sorbent

Sorbent materials must have high selectivity/affinity with a capacity to enable sharp separation and minimize the amount of sorbent required to make a suitable commercial product. The sorbent should have favorable kinetics and transport properties for the rapid adsorption of target solutes, chemical and thermal stability, low solubility in the contacting fluid, and high mechanical strength to prevent crushing or erosion.

In a sorbent cartridge used for clinical purposes, the material must allow free flow of blood or plasma (fluid phase) and easy filling and emptying of the packed bed. Other requirements are high resistance to fouling to permit long cartridge life span and maximal biocompatibility with no tendency to promote undesirable chemical reactions or side effects.

Mechanism of Solute Adsorption in Porous Media

Different steps and mechanisms can be identified in the process of solute adsorption onto a porous material: (a) external (inter-phase) mass transfer of the solute by convection from the bulk fluid and by diffusion through a thin film or boundary layer, to the outer surface of the sorbent; (b) internal (intra-phase) mass transfer of the solute by convection from the outer surface of the sorbent into the internal porous structure; and (c) surface diffusion along the surface of the internal pores and adsorption of the solute onto the porous surface (Fig. 2). The adsorption mechanism involves physiochemical forces of different nature [7–9].

The interphase mass transfer is a crucial step since it brings the solution (fluid phase) and the molecules to be removed in contact with the sorbent. The cartridge in which the sorbent is contained must promote uniform distribution of internal flow of the fluid phase (plasma or whole blood). Uniform flow distribution profiles are generally obtained using granules or spherical beads of equal size. Packing density between 40 and 60% is considered optimal to prevent preferential channeling of the flow with undesired loss of performance. Any type of channeling phenomenon may affect the quantity of solute adsorbed per unit of sorbent and influence the saturation process of the unit.

Since blood is a non-Newtonian fluid, accurate analysis of the flow distribution in different conditions of flow and viscosity should be made. Flow distribution in packed beds can be theoretically modeled using equations of physical chemistry and transport. The packing structure is usually complex, and the resulting flow pattern is complicated. There are tortuous paths through the interstitial space of the bed, which consists of channels of various diameters (inter-particle porosity). The packed bed can simulate a bundle of tortuous capillary tubes. In well-packed beds with relatively constant inter-particle porosity, the variation of flow velocity among individual channels is relatively small. However, if packing is not homogeneous, channels of different size can be present with significant variation of fluid phase velocity, leading potentially to blood stagnation due to high resistance in areas with small-diameter channels and consequent clotting. On the contrary, areas having large diameter channels offer relatively little resistance to flow and the undesirable phenomenon of preferential flow channeling may result, with poor utilization of the sorbent potential, reduction in adsorption performance and rapid saturation of the unit.

The internal mass transfer (intra-phase) can be seen as a convective transport of the solute through the structure of the sorbent due to flow of the fluid phase inside the sorbent particle. This once again depends on the packing density, the pressure gradient, and the permeability coefficient of the particle. Often this mechanism is far from being optimized and the sorbent is generally utilized only in minimal part due to insufficient permeation of the bulk solution into the structure of the particle.

The physiochemical mechanisms regulating the surface adsorption are multiple. Once the molecule is brought to the surface of the sorbent, different chemical and physical mechanisms are involved:

- van der Waals forces are generated by the interaction between electrons of one molecule and the nucleus of another molecule; these are weak and generally reversible.

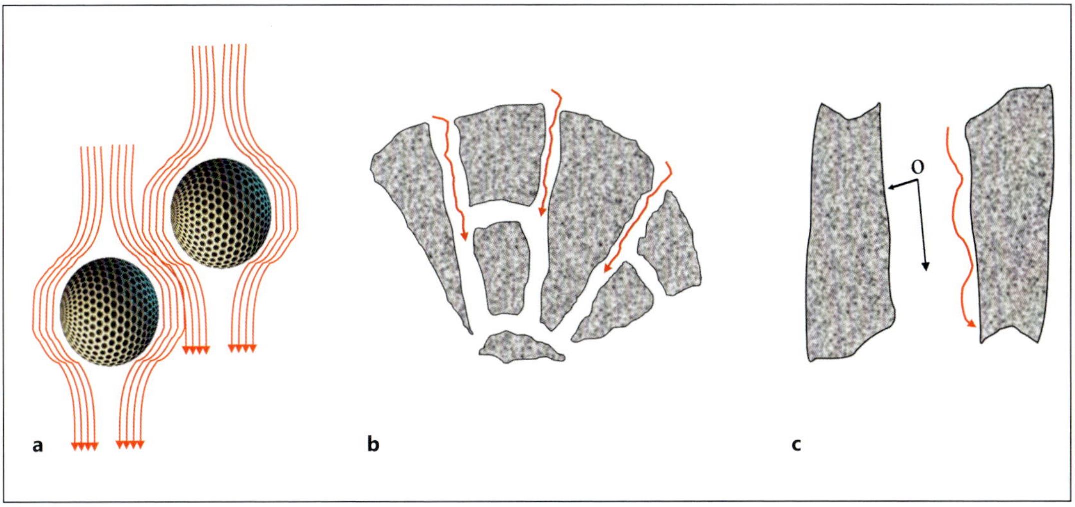

Fig. 2. Mechanisms of mass transport from the bulk solution to the sorbent surface. **a** External (interphase) mass transfer of the solute by convection from the bulk fluid by diffusion through a thin film or boundary layer to the outer surface of the sorbent. **b** Internal (intraphase) mass transfer of the solute by pore convection from the outer surface of the adsorbent to the inner surface of the internal porous structure. **c** Surface diffusion along the porous surface and adsorption of the solute onto the porous surface.

- Ionic bonds are generated by electrostatic attraction between positively charged and negatively charged ions; these are typical of ion exchange resins.
- Hydrophobic bonds represent strong binding forces that are generated by the hydrophobic affinity of the sorbent and solute molecules (Fig. 2).

Biocompatibility of Sorbents

The biocompatibility of a system utilizing sorbents for extracorporeal therapies should be studied considering the different aspects. First, the sorbent must be resistant and have sufficient mechanical strength to prevent cracking of the solid component, with release of micro-particles and fragments to the systemic circulation. To further prevent this unwanted effect, cartridges are provided with a screen that allows free passage of blood but retains particles or their fragments. A derivative measure of biocompatibility is given in clinical practice by the continuous measure of end-to-end pressure drop in the unit throughout the treatment. Fouling of screens due to cell or albumin adhesion may result in increased resistance to flow and thus increased pressure drop inside the cartridge. Accelerated clotting of the unit will also cause a sudden increase in end-to-end pressure drop.

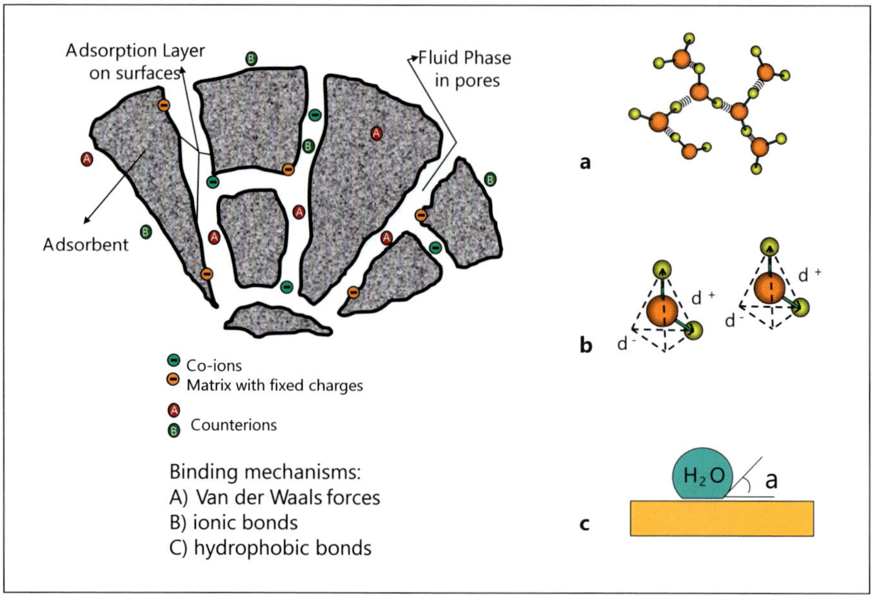

Fig. 3. Left: physicochemical mechanisms regulating molecular surface adsorption. Right: once the molecule is brought to the surface of the sorbent, different chemical and physical forces play the final role: (**a**) van der Waals forces generated by the interaction between electrons of one molecule and the nucleus of another molecule (weak and generally reversible); (**b**) ionic bonds generated by electrostatic attraction between positively charged and negatively charged ions (typical of ion exchange resins); (**c**) hydrophobic bonds generated by the hydrophobic affinity of the sorbent and solute molecules.

The second aspect is the intrinsic structure of the sorbent material. The inner surface of the sorbent should be compatible with blood to avoid cell and protein deposition that may occupy the adsorption sites and impair the sorbent capacity. When the material is intended for direct contact with blood, biocompatibility should be further directed towards preventing unwanted reactions in circulating blood (from complement activation to cytokine release), leukopenia, thrombocytopenia, development of antibodies, and significant adsorption of albumin. All these effects can be mitigated by coating the surface of the granules or fibers with a biocompatible material such as polysulfone. In this case, however, the coating may render the sorbent less efficient because the intra-phase component of the transport may be negatively affected. The coating acts as a size exclusion barrier and prevents larger solutes from reaching the intra-particle site of adsorption.

To obviate the need for coating the sorbent, some techniques separate plasma from cells and circulate cell-free plasma through the sorbent bed, avoiding direct contact with cells. Downstream in the circuit, blood is reconstituted by mixing purified plasma with cells [10]. In some cases, only plasma water ultrafiltrate is regenerated by exposing it to the sorbent bed and subsequently reinfusing it downstream into the circuit.

Rationale for the Use of Adsorption in Clinical Settings

For many years, the use of sorbents was mostly proposed for chronic hemodialysis patients to remove molecules that were not easily removed by hemodialysis. Sorbents were also indicated in case of drug intoxication and poisoning where toxin removal had to be obtained rapidly and efficiently. More recently, a rationale for the use of sorbents in critical illness, sepsis, and acute kidney injury has emerged due to the proposed humoral pathogenesis of these disorders. Assuming there is a humoral disorder with pathologic circulating molecules (e.g., damage-associated molecular patterns; pathogen-associated molecular patterns), extracorporeal therapies designed to remove these molecules would offer potential benefits [11]. There is a possibility to employ selective sorbents to target specific molecules. In clinical practice, however, although the use of sorbents may offer some interesting advantages, all other aspects such as hemocompatibility, unwanted solute losses, or alteration of anticoagulation requirements in the extracorporeal circuit should be considered [11].

Most Common Modalities of Utilization of Sorbents

Sorbent-based techniques have been applied to the management of both acute kidney injury and chronic kidney disease [10–12].

Hemoperfusion
HP is a technique in which the sorbent is placed in direct contact with blood in an extracorporeal circulation [8]. A peristaltic pump via blood lines circulates blood through the sorbent cartridge. HP has a circuit simpler than hemodialysis, but requires adequate anticoagulation and a very biocompatible sorbent since there is a direct contact between blood and sorbent material. Charcoal has a high adsorbing capacity, especially for relatively hydrophobic, low molecular weight solutes that are retained in case of kidney or liver failure. Direct contact of blood with charcoal in the absence of a biocompatible coating, however, is not advised. On the contrary, a coating may markedly reduce the adsorptive capacity of the carbon due to the size exclusion effect of the coating layer. More recently, synthetic polymers with remarkable capacity of adsorption have been made avail-

able for clinical HP. The pores on the surface of the granules have been enlarged such that size exclusion has become a minor issue. Due to these recent advances, sorbent units are today available for direct HP and have been demonstrated to be quite efficient in removing poisons, bilirubin, cytokines or even endotoxin.

Coupled Plasma Filtration-Adsorption

Continuous plasma filtration-adsorption (CPFA) is a modality of blood purification in which plasma is separated from the whole blood by a plasma filter and circulated in a sorbent cartridge. After the sorbent unit, plasma is returned to the blood circuit and the reconstituted whole blood undergoes hemofiltration or hemodialysis. Today, specific ready-to-use disposable kits are prepared for easy application of the technique [10]. The aim of this approach is to attempt to achieve adequate removal of molecules, typically hydrophobic in nature, that are not removed appreciably by other hemofiltration or hemodialysis techniques. The advantage is to exclude blood cells from contact with the sorbent and to re-infuse endogenous plasma after non-selective simultaneous removal of different sepsis-associated mediators, without the need of donor plasma. This technique has been mostly used in septic patients [13] showing specific advantages of blood purification, restoration of hemodynamics, and immunomodulation.

The results obtained in clinical practice were in fact the basis to formulate the "peak concentration hypothesis" and to offer a possible explanation of the beneficial effects of sorbents in septic patients [14]. The unselective but continuous diminution of the peak concentrations of both pro- and anti-inflammatory mediators may in fact lead to a type of immune-modulation with partial restoration of immuno-homeostasis.

Sorbent-Based Adjunctive Therapies for Sepsis

This wider approach to the concept of blood purification opens new perspectives in a revisited strategy for the application of sorbents and extracorporeal therapies, especially in the area of sepsis [14]. The cellular and humoral responses of the host to bacterial invasion result in a series of symptoms and organ derangements, which are mediated by the presence of chemical substances found in the plasma and tissue. Continuous renal replacement therapies (CRRT) have gained increased popularity for their ability to facilitate the removal of excess fluid and waste products in septic patients with acute kidney injury. However, removal rates and clearances of different proinflammatory cytokines (IL-1, TNF) and lipid mediators (PAF) are hindered by insufficient membrane permeability. To overcome such limitations, high volume hemofiltration and use of high cut-off membranes have been proposed. The latter are still under investigation, both for potential benefits and also possible drawbacks (e.g., excessive leakage of albumin). Plasma filtration techniques (plasmapheresis and plasma exchange) have shown an increase in TNF clearance by 2 orders of magnitude relative to standard CRRT and have demonstrated some possible survival benefits in septic animals. However, plasmapheresis cannot be considered a routine therapy, due to technical complexity and high costs.

In addition to CPFA, the following are commonly applied adsorptive therapies for sepsis:

CytoSorb™

CytoSorb™ is the first-in-class therapy specifically CE marked as an extracorporeal cytokine cartridge in the European Union. Its use is broadly indicated for clinical conditions in which plasma cytokine concentrations are elevated. The unit contains a biocompatible, highly porous polymer bead designed to capture and adsorb cytokines in the ~10–50 kDa range [15]. The goal is to reduce toxic cytokine levels to prevent or mitigate organ failure and immune suppression, thereby improving the clinical outcome. Clinical trials are underway in septic patients to establish the real clinical influence of this device on short- and long-term outcomes. Another use of this cartridge is the intended removal of cytokines from the extracorporeal circulation in the case of extracorporeal membrane oxygenation applied in the context of cardiac surgery.

Polymyxin-B-HP

In the area of sorbents applied to blood purification, the technique of polymyxin-B (PMX) HP for endotoxin removal has gained important evidence and results [16, 17]. This direct HP technique utilizes a unit in which polystyrene-based fibers are functionalized with covalently bound PMX. This compound is a potent antibiotic that acts as an avid scavenger of circulating lipopolysaccharide S, the major component of bacterial endotoxin. Because endotoxin is the trigger of many humoral and cellular reactions in sepsis leading to organ damage and dysfunction, there is a clear rationale for endotoxin removal in sepsis.

From a historical perspective, the PMX device (Toraymyxin: Toray Industries, Japan) for extracorporeal removal of endotoxin was introduced in Japan several years ago and is intended to represent an adjuvant sepsis therapy. The treatment is particularly indicated in septic shock from Gram-bacteria with high levels of circulating endotoxin assessed by specific assays. Due to the high affinity of PMX to endotoxin, the rationale of the extracorporeal therapy is to prevent the evolution of the biological cascade of sepsis. Although studies have often produced confusing or even conflicting results [18–22], PMX is routinely used in Japan since 1995 and more than 50,000 septic patients have been treated to date.

A group of investigators was coordinated by 2 centers including our hospital in the study called Early Use of Polymyxin-B Hemoperfusion in Abdominal Sepsis (EUPHAS). This multicenter randomized controlled study provided the first evidence that PMX can achieve a significant reduction in mortality for patients with abdominal septic shock, after a careful analysis of the literature had reported a similar retrospective result [20].

In 2010, the EUPHAS2 project created a registry with the purpose of recording data from critically ill septic patients affected by severe sepsis and septic shock, and treated with PMX-based direct HP (PMX-DHP) for endotoxin removal [21]. The aim of the registry was to characterize the application of PMX-DHP in daily clinical practice. The registry involved 46 European and 11 Asian hospitals, collecting retrospective data of 357 patients (297 in Europe and 60 in Asia) from 35 centers between January 2010 and December 2014. Finally, EUPHRATES (evaluating the use of polymyxin B hemoperfusion in a randomized controlled trial of adults treated for endotoxemia and septic shock) was conducted in Canada and North America [22]. The initial results from the trial have been announced recently and further analysis of the potentially beneficial effects of PMX-DHP in specific subgroups of septic patients with high endotoxin levels is ongoing.

Novel Applications of Sorbents

Lixelle Beta 2-Microglobulin Apheresis Column

The Lixelle column (Kaneka Co., Tokyo, Japan) contains 350 mL of porous cellulose adsorbent beads (diameter ~460 μm) to which are attached a ligand containing a hydrophobic hexadecyl group [23]. Peptides and proteins of molecular weight less than 20,000 Da are able to permeate the bead pores and attach to the ligand by hydrophobic interactions. The US FDA approved the column by a humanitarian device exemption in 2015 with an indication of dialysis-related amyloidosis in end-stage renal disease (ESRD) patients, and it is currently being evaluated in a post-approval trial.

Virus Removal by Lectin Affinity Plasmapheresis

The hemopurifier device (Aethlon Medical, San Diego, CA, USA) is a plasma filter which exploits the known ability of certain lectin-based compounds to bind viral particles. In this device, a resin to which an agglutinin ligand is attached acts as an affinity matrix for the binding of virions and viral glycoproteins. The resin comprises the non-blood compartment of the plasma filter and binds viral particles in the plasma filtrate produced in the proximal third of the filter. Based on the transmembrane pressure profile, the filtrate re-enters the blood compartment at a more distal stage through a Starling's flow mechanism. Successful treatment with this device (in series with a CRRT filter) of a critically ill patient infected with the Ebola virus has been reported recently [24].

HP for Sepsis-Induced Acute Lung Injury

The HA300 device (Jaffron Biomedical, Zhuhai, China) is a novel HP column containing neutral macroporous beads. With an approximate pore size distribution corresponding to a molecular weight range of 10–60 kDa, it is well suited for the removal of many inflammatory mediators that are relevant in sepsis. The device has been evaluated prospectively in 46 patients with acute lung injury related to extrapulmonary sepsis [25]. Patients in the intervention group ($n = 25$) received treatment on 3 consecutive days plus standard care. In comparison to the control group, hemodynamic and respiratory parameters were significantly improved at day 7. Moreover, both ICU and 28-day mortality were significantly lower in the HP group.

Sorbent Technology in Wearable Artificial Kidney Devices

Finally, due to limitations of currently available ESRD therapies, substantial interest now exists in applying the principles of miniaturization to the development of wearable/implantable dialysis devices [26]. In a version developed by Gura et al.[27], spent dialysate is regenerated using a sorbent-based system that includes urease, zirconium phosphate, hydrous zirconium oxide, and activated carbon. It is expected that continued advancements in this area will occur in the future.

Conclusions

Although adsorbent materials have been routinely employed in clinical medicine for poisoning and intoxication, their use for other clinical applications has been relatively rare. An improved knowledge of the manufacturing processes and the possibility of designing new sorbents with improved characteristics of biocompatibility offer new opportunities with enormous potential. A new series of studies has demonstrated the feasibility, safety, and clinical benefits of using sorbents alone or in combination with other techniques. Therefore, sorbent technologies may assume a larger role for a variety of clinical disorders in the future.

References

1 Ronco C, Bordoni V, Levin NW: Adsorbents: from basic structure to clinical application. Contrib Nephrol 2002;137:158–164.

2 Winchester JF, Ronco C, Brady JA, et al: Adsorbent augmented dialysis: minor addition or major advance in therapy? Blood Purif 2001;19:255–259.

3 Samtleben W, Gurland HJ, Lysaght MJ, Winchester JF: Plasma exchange and hemoperfusion; in Jacobs C, Kjellstrand CM, Koch KM, Winchester JF (eds): Replacement of Renal Function by Dialysis. Boston, Kluwer Academic, 1996, pp 472–500.

4 Ronco C, Brendolan A, Winchester JF, et al: First clinical experience with an adjunctive hemoperfusion device designed specifically to remove beta(2)-microglobulin in hemodialysis. Blood Purif 2001;19:260–263.

5 Clark WR, Rocha E, Ronco C: Solute removal by hollow-fiber dialyzers. Contrib Nephrol 2007;158:20–33.

6 Donati G, La Manna G, Cianciolo G, et al: Extracorporeal detoxification for hepatic failure using molecular adsorbent recirculating system: depurative efficiency and clinical results in a long-term follow-up. Artif Organs 2014;38:125–134.

7 La Greca G, Brendolan A, Ghezzi PM, et al: The concept of sorbent in hemodialysis. Int J Artif Organs 1998;21:303–308.

8 Tetta C, Bellomo R, Brendolan A, et al: Use of adsorptive mechanisms in continuous renal replacement therapies in the critically ill. Kidney Int Suppl 1999;56(suppl 72):S15–S19.

9 Ronco C, Ghezzi PM, Morris A, et al: Blood flow distribution in sorbent beds: analysis of a new sorbent device for hemoperfusion. Int J Artif Organs 2000;23:125–130.

10 Tetta C, Cavaillon JM, Camussi G, et al: Continuous plasma filtration coupled with sorbents. Kidney Int Suppl 1998;53(suppl 66):S186–S189.

11 Winchester JF, Kellum JA, Ronco C, et al: Sorbents in acute renal failure and the systemic inflammatory response syndrome. Blood Purif 2003;21:79–84.

12 Winchester JF, Ronco C, Brady JA, et al: Rationale for combined hemoperfusion/hemodialysis in uremia. Contrib Nephrol 2001; 133:174–179.

13 Ronco C, Brendolan A, Dan M, et al: Adsorption in sepsis. Kidney Int 2000;58(suppl 76):S148–S155.

14 Ronco C, Bonello M, Bordoni V, Ricci Z, D'Intini V, Bellomo R, Levin NW: Extracorporeal therapies in non-renal disease: treatment of sepsis and the peak concentration hypothesis. Blood Purif 2004;22:164–174.

15 Taniguchi T: Cytokine adsorbing columns. Contrib Nephrol 2010;166:134–141.

16 Nemoto H, Nakamoto H, Okada H, et al: Newly developed immobilized polymyxin B fibers improve the survival of patients with sepsis. Blood Purif 2001;19:361–368; discussion 368–369.

17 Shoji H: Extracorporeal endotoxin removal for the treatment of sepsis: endotoxin adsorption cartridge (Toraymyxin). Ther Apher Dial 2003;7:108–114.

18 Antonelli M, Ronco C: Polymyxin B hemoperfusion in sepsis: growing body of evidence and occasional conflicting results. Blood Purif 2015;39:I–II.

19 Payen DM, Guilhot J, Launey Y, et al; ABDOMIX Group: Early use of polymyxin B hemoperfusion in patients with septic shock due to peritonitis: a multicenter randomized control trial. Intensive Care Med 2015;41:975–984.

20 Cruz DN, Antonelli M, Fumagalli R, et al: Early use of polymyxin B hemoperfusion in abdominal septic shock: the EUPHAS randomized controlled trial. JAMA 2009;301: 2445–2452.

21 Martin EL, Cruz DN, Monti G, Casella G, Vesconi S, Ranieri VM, Ronco C, Antonelli M: Endotoxin removal: how far from the evidence? The EUPHAS 2 project. Contrib Nephrol 2010;167:119–125.

22 Klein DJ, Foster D, Schorr CA, Kazempour K, Walker PM, Dellinger RP: The EUPHRATES trial (evaluating the use of polymyxin B hemoperfusion in a randomized controlled trial of adults treated for endotoxemia and septic shock): study protocol for a randomized controlled trial. Trials 2014;15:218.

23 Abe T, Uchita K, Orita H, et al: Effect of beta(2)-microglobulin adsorption column on dialysis-related amyloidosis. Kidney Int 2003; 64:1522–1528.

24 Büttner S, Koch B, Dolnik O, et al: Extracorporeal virus elimination for the treatment of severe Ebola virus disease – first experience with lectin affinity plasmapheresis. Blood Purif 2014;38:286–291.
25 Huang Z, Wang SR, Yang ZL, Liu JY: Effect on extrapulmonary sepsis-induced acute lung injury by hemoperfusion with neutral microporous resin column. Ther Apher Dial 2013; 17:454–461.
26 Ronco C, Davenport A, Gura V: The future of the artificial kidney: Moving towards wearable and miniaturized devices. Nefrologia 2011;31:9–16.
27 Gura V, Macy AS, Beizai M, Ezon C, Golper TA: Technical breakthroughs in the wearable artificial kidney (WAK). Clin J Am Soc Nephrol 2009;4:1441–1448.

William R. Clark, MD
Purdue University School of Chemical Engineering
480 Stadium Mall Drive, Forney Hall Room G051
West Lafayette, IN 47907 (USA)
E-Mail clarkw@purdue.edu

La Manna G, Ronco C (eds): Current Perspectives in Kidney Diseases.
Contrib Nephrol. Basel, Karger, 2017, vol 190, pp 58–70 (DOI: 10.1159/000468914)

Development of the New Kibou® Equipment for Continuous Renal Replacement Therapy from Scratch to the Final Configuration

Mauro Neri[a] · Anna Lorenzin[a] · Alessandra Brendolan[a, b] · Francesco Garzotto[a, b] · Fiorenza Ferrari[a, c] · Silvia De Rosa[a, c] · Raffaele Bonato[c] · Gianluca Villa[a] · Simona Bazzano[a] · Fiorella D'Ippoliti[a] · Zaccaria Ricci[d] · Gaetano La Manna[e] · Claudio Ronco[a, b]

[a]International Renal Research Institute of Vicenza (IRRIV), [b]Department of Nephrology, Dialysis and Transplantation, and [c]Department of Intensive Care Unit, San Bortolo Hospital, Vicenza, [d]Department of Cardiology and Cardiac Surgery, Pediatric Cardiac Intensive Care Unit, Bambino Gesù Children's Hospital, IRCCS Rome, Rome, and [e]Department of Experimental, Diagnostic and Specialty Medicine, Nephrology, Dialysis and Transplantation Unit, St. Orsola Hospital, University of Bologna, Bologna, Italy

Abstract

A new technology has recently appeared in the area of extracorporeal therapies for critically ill patients with acute kidney injury. The International Renal Research Institute of Vicenza was involved from the beginning in the development of a new continuous renal replacement therapy (CRRT) equipment with peculiar characteristics. We report the overall experience from design of the new machine to its in vitro and in vivo testing. Kibou® (Asahi Kasei Kuraray Medical Co., Ltd., Tokyo, Japan) is a new multifunctional machine designed for delivering RRT. Kibou® carries out many features of the fourth generation CRRT machines including the possibility of a dynamic prescription and reduction of nursing workload. We describe our first experience with this new device, focusing on several usability and performance parameters. A specific in vitro protocol was designed to analyze the various characteristics and accuracy of performance of the machine. Furthermore, a preliminary in vivo alpha trial with 12 CRRT sessions was performed to test, characterize and evaluate the machine in terms of usability, flexibility and reliability. The in vitro evaluation confirmed an adequate design and a good usability of the machine with accurate delivery of prescribed parameters. No adverse events were observed during the in vivo test that confirmed usability and safety together with accuracy of treatment delivery in different modalities. In general, the machine was rated by physicians and nurses involved in the evaluation as practical and easy to use, although a specific training is required to familiarize with the equipment. A large-scale multicenter beta trial is required to confirm the results reported in this preliminary evaluation in terms of safety, accuracy and performance of Kibou®.

Introduction

Acute kidney injury (AKI) is a disease that frequently occurs in intensive care patients, with high mortality rates. This clinical entity has required, over the years, a progressive scientific effort to develop adequate technology for providing safe and reliable treatments.

Continuous renal replacement therapy (CRRT), applied through dedicated machines, is a first-line therapy for the treatment of AKI. The modern history of CRRT is characterized by the development of machines designed specifically for acute renal replacement in intensive care patients. Although in the past these machines were basically derived from hemodialysis blood modules, in the last years they are designed exclusively as self-standing units for CRRT.

The evolution of CRRT machines has seen different eras since the late 1970s [1]. The first CRRT devices appeared in 1977 [2] when continuous arteriovenous hemofiltration (CAVH) was performed. CAVH presented important advantages concerning hemodynamic stability, control of circulating volume and possibility of nutritional support, but the need for arterial cannulation and limited solute clearance represented serious shortcomings. The development of veno-venous CRRT systems in the late 1980s represented a milestone, and formed the basis for modern therapies. At that time, CRRT equipment borrowed some components from chronic hemodialysis machines such as blood pump and ancillary pumps to control volumes of dialysate and/or replacement fluid. Fluid delivery systems and net ultrafiltration control mechanisms were implemented allowing acceptable accuracy. The third generation of machines [3] guaranteed greater blood/fluid flow rate capabilities, additional functionalities and components, and safety features today recommended in every modern equipment. The fourth and latest generation of machines for CRRT, available or to be launched next in the market [4, 5], represents the progressive and continuous evolution of the past forty years of research and development in this field (Fig. 1). The main target of the new generation of machines is an accurate delivery of the prescribed dialysis dose, high safety and easy utilization with a reliable performance of the therapy [6]. In particular, the reduction of

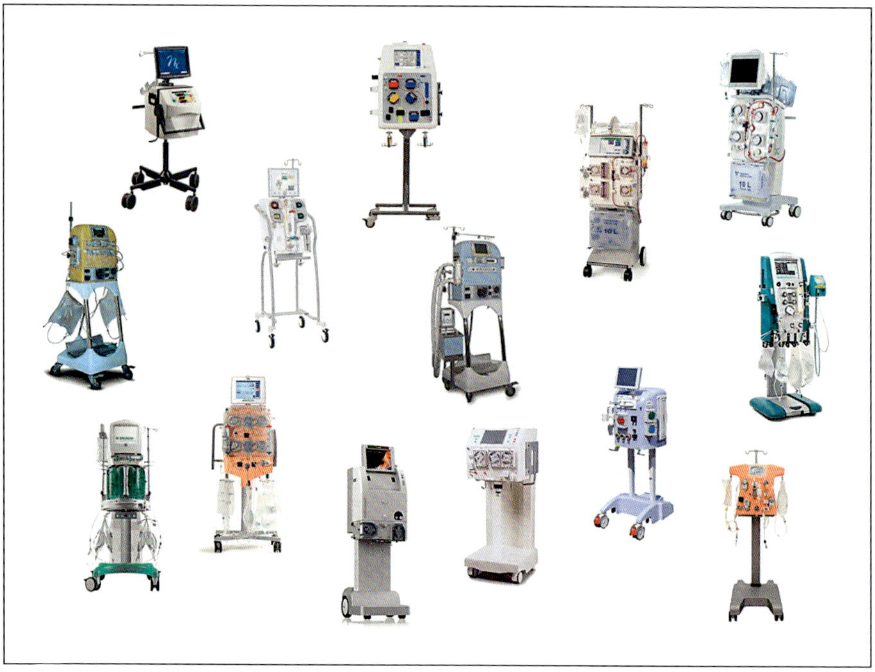

Fig. 1. CRRT machines from the last 20 years until today [1].

actual dose due to downtime, optimization of alarms management, making the nursing workload lighter and the development of safe protocols for new anticoagulation therapies (e.g., citrate) have been identified as the most critical points.

As a common line, these new devices allow us to perform almost all the available CRRT techniques and are equipped with a groundbreaking touch screen and friendly user interface, like the last generation mobile phones. Clear alarm signals and warnings, together with important treatment parameters like pressure trends, are now clearly displayed on the screen. To provide clinicians with individualized CRRT treatment data, machines have also been implemented and equipped with practical tools to easily download treatment parameters [7].

As in previous occasions, our institute has been recently involved in the development and testing of some of the new generation CRRT machines. Kibou® (Asahi Kasei Kuraray Medical Co., Ltd., Tokyo, Japan) is an example of new technology developed from scratch to a final equipment. "Kibou," a Japanese word meaning "hope," is a CE marked, multifunctional, automatic CRRT machine designed to deliver and monitor continuous blood purification treatments. In this chapter, we will describe the features and characteristics of the

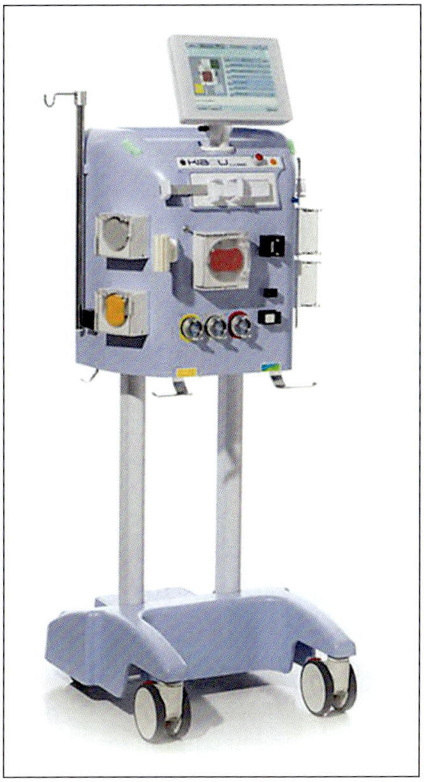

Fig. 2. Layout of Kibou® machine.

machine according to our initial design specifications, highlighting peculiarities and new aspects. Our in vitro and first in vivo evaluation is reported with specific user ratings including clinical, practical and technical aspects.

Kibou® Design: Features and Characteristics

The Kibou® machine, displayed in Figure 2, is a CRRT platform featuring many characteristics of the fourth generation machines, previously mentioned. Frontally looking, the hardware layout is divided into three panels: the right one is completely dedicated to fresh solutions (dialysate and replacement fluid) necessary to purify blood by convection or diffusion, the central one is dedicate to blood circulation and all safety system sensors and the left part concerns the "dirty" effluent and citrate infusion components (pump and scale).

The machine features a 12-inches, full color touch screen monitor, with function keys required for each phase and process. In total, Kibou® is equipped with 5 peristaltic pumps: blood (range 1–400 mL/min), dialysate/replacement

Table 1. Modalities and setting parameters performable with Kibou®

CRRT treatment	Adult		Pediatric		Replacement fluid			Dialysate fluid	
	heparin	citrate-calcium	heparin	citrate-calcium	pre	post	pre and post	co-current	counter-current
SCUF	✓	✓	✓	✓	x	x	x	x	x
CVVH	✓	✓	✓	✓	✓	✓	✓	x	x
CVVHD	✓	✓	✓	✓	x	x	x	✓	✓
CVVHDF	✓	✓	✓	✓	x	✓	x	✓	✓
TPE	✓	✓	✓	✓	x	x	x	x	x

CRRT, continuous renal replacement therapy; SCUF, slow continuous ultrafiltration; CVVH, continuous veno-venous hemofiltration; CVVHD, continuous veno-venous hemodialysis; CVVHDF, continuous veno-venous hemodiafiltration; TPE, therapeutic plasma exchange.

pre-filter (range 10–10,000 mL/h), replacement post-filter (range 10–10,000 mL/h), effluent (range 10–12,000 mL/h) and citrate (10–600 mL/h). One syringe pump is dedicated to heparin or calcium infusion.

Three scales are allocated on the bottom of the machine. Scales for replacement fluid/dialysate and effluent can manage a maximum load of 15 kg. On the citrate one, the maximum load is 2 kg.

CRRT modalities and plasma exchange treatment for both adults and pediatrics are available (Table 1). In continuous veno-venous hemodialysis (CVVHD) and continuous veno-venous hemodiafiltration (CVVHDF), both dialysate counter-current and co-current configurations are possible. Heparin or citrate-calcium anticoagulation therapies can be prescribed and delivered.

The disposable tubing set of the machine is the same for all CRRT treatments: slow continuous ultrafiltration (SCUF), continuous veno-venous hemofiltration (CVVH), CVVHD, CVVHDF. Citrate anticoagulation therapy, pediatric therapy and therapeutic plasma exchange (TPE) are performed with dedicated tubing sets. The machine can be set with the following disposable configurations:

- Adult CRRT kit with heparin as anticoagulation therapy
- Adult CRRT kit with citrate-calcium as anticoagulation therapy
- Pediatric CRRT kit with heparin as anticoagulation therapy
- Pediatric CRRT kit with citrate-calcium as anticoagulation therapy
- Adult TPE kit with heparin as anticoagulation therapy
- Pediatric TPE kit with heparin as anticoagulation therapy.

The pediatric CRRT kit has a priming volume of 80 mL (adult one is 300 mL) and can be applied for patients whose weight is higher than 10 kg.

In case of citrate anticoagulation, several solutions with different concentrations of citrate and calcium can be used:

- Citrate:
  - Anticoagulant citrate dextrose solution: 112.9 mmol/L
  - Trisodium citrate: 136 mmol/L.
- Calcium:
  - Calcium chloride: 4.56 mmol/10 mL
  - Calcium gluconate: 2.27 mmol/10 mL
  - High concentration solution: 500 mmol/L.

The hardware is designed to be used with any type of dialyzer/filter and dialysate/replacement fluid bag. During the setup phase of the machine, dry or wet dialyzer can be set. Wet filters can be used to improve the air removal process during the priming phase and to prevent type 1 reaction occasionally observed with new dry dialyzers [8]. If diffusive therapy is performed, standard counter- or co-current configuration of dialysate flow can be set. In a practical setting, priming, mounting and use of CRRT circuit are easier in co-current dialysate flow because gas and air bubbles will be removed from the membrane more efficiently [9]. This setting does not reduce significantly the treatment efficiency if blood and dialysate flows are optimized.

The therapy prescription can be set in the machine in 2 ways:

- Setting the flows (or volumes) of purifying fluids
- Setting the dose intended to achieve.

In the second case, the machine will automatically output the flows based on set parameters necessary for the calculation.

In vitro and In vivo Evaluation

Based on the traditional experience of International Renal Research Institute of Vicenza (IRRIV), in vitro and first in vivo evaluation of Kibou® CRRT machine has been conducted with the classic approach of usability and accuracy test.

Staff Involved

Participants for this device evaluation were nephrologists, intensivists, dialysis nurses, intensive care nurses and engineers working in the research center.

In vitro Testing

The in vitro evaluation protocol of the machine has been divided into 5 subsections as:

- General evaluation
- Hardware evaluation
- Software evaluation (bugs, graphical user interface [GUI])

- Alarm simulation phase
- Fluid balance error.

General Evaluation
The general perception of the machine was quantified through the administration of a survey to each component of the evaluating staff at the end of the in vitro testing. Aspects of functionality, usability, feasibility during all phases of a theoretical CRRT session was judged based on a score between 1 and 5. The purpose was to bring out peculiarities, innovative features, weaker points and issues.

Hardware Evaluation
Not only was the machine in its entirety evaluated, but also each hardware component (display, tubing set, pumps, warmer, scales, pressure transducers, exc.). Evaluation criteria were the same as before.

Software Evaluation (Bugs, GUI)
Functionalities, usability, ease of understanding and accuracy of displayed parameters of GUI were evaluated. In this phase, even software bugs were tracked, highlighted and solved.

Alarm Simulation Phase
In this phase, alarming situations were induced by altering pressures, flows and circuit setup to evaluate the safety of the machine. We checked the type of alarm, intervals of detection and possible alterations of all components. A strict protocol based on the alarms list reported in the user manual was followed.

Fluid Balance Error
During this phase, sessions of 6 and 24 h with physiological saline solution and low fat milk were performed to evaluate the accuracy of fluid balance system of the machine, including scales and software. Measurements after 1, 3, 6 and 24 h were considered. Different modalities and flows were applied. Direct volume measurement of each "compartment" (bag simulating the patient, dialysate bag, replacement bag and effluent bag) was performed using a high-resolution scale. Particular attention was paid on the value of net ultrafiltration volume at specific time points. Comparison between values displayed on the screen and values obtained by weighing bags was considered to estimate the relative fluid balance error.

In vivo Testing
After a deep and long in vitro analysis aiming to technically improve the machine, a preliminary in vivo "alpha test" was performed in our semi-intensive care unit

of the Department of Nephrology, Dialysis and Transplantation in Vicenza, Italy. The test was performed on nephrology inpatients requiring a short-time session of RRT. We performed 12 RRT treatments in 9 patients. Although Kibou® was already a commercially available CE marked device, we obtained the consent by our internal review board and informed consent from each treated patient.

Staff Training
Nephrologists, intensivists, nurses and engineers involved in the study received a specific training program on the machine.

Prescription
The prescription of CRRT modalities and treatment parameters to be set in the machine was completely left to the discretion of the physician in charge of the patient, based on the specific clinical need.

RRT Disposables and Tubing Set
The tubing set, applied for all treatments, was the adult kit with heparin as an anticoagulant. The dialyzer applied for this preliminary alpha trial was Cureflo 1.3 (Asahi Kasei Kuraray Medical Co., Ltd., Tokyo, Japan). Bags for replacement and dialysate solutions were Duosol™ Bicarb 35 Dialysate K2/Ca3 (B. Braun Medical Inc., Bethlehem, PA, USA).

Collection of Treatment Data and Parameters
Technical data and parameters of the treatment were collected manually and extracted from the device through a USB key. No clinical data were collected since the clinical evaluation of the machine was not included in the aims of the test.

Statistical Analysis
Descriptive data of in vivo treatments are expressed as mean ± SD. Excel software (Microsoft, Redmond, WA, USA) was used for the analysis.

Evaluation Results

In vitro Testing
In general, Kibou® is considered a small and compact CRRT machine. Based on assigned scores, it allows for easy handling and logistical movement (3.3/5). Visual and acoustic signals for alarms announcement are clear to perceive, especially in a noisy environment like ICU departments (3.9/5). Nevertheless, the steel arm for hanging priming bags could probably be taller and equipped on the

top with a further LED for an immediate perception of the alarm. The display is quite small (2.7/5), but its dimensions are adequate for an easy rotation on the vertical axis (3.5). The space of the frontal layout is perceived as well organized, especially during the preparation phase (3.6).

Regarding the evaluation of hardware components and disposable, peristaltic pumps (3.3/5), pressure sensors (3.5/5) and all other safety devices allocated in the machine (air bubbles detector, blood leak detector, safety automatic clamp, exc.) received good scores, both in terms of positioning and circuit handling. The installation phase of the disposable part of the tubing set on the heater of the machine was considered a bit cumbersome (1.9/5). The automatic adjustment of the blood level in the air removal chamber in the return line, achieved by 2 ultrasound probes, was considered very useful. However, the shape of the disposable chamber and hemodynamics inside it, may lead to some stagnation of blood and increase risks of clotting.

Dialysate/replacement and effluent scales received a good mark (3.0/5), especially from nurses, because the maximum weight potentially to load (15 kg) was considered practical, potentially reducing the frequency of bags change procedures.

The pediatric blood pump segment has a lower stroke volume than the adult one. A 0.15 m^2 dialyzer was used. Blood flow was accurate for values of 50 mL/min (increments of blood flow rate is 1 mL/min from 1 to 30 mL/min, 5 mL/min from 30 to 400 mL/min). However, blood and effluent pumps work in a discontinuous way for very low flows (1–30 mL/min).

Also, software and graphic user interface were evaluated and considered immediate, comprehensive (3.2/5) and with a good setup and organization on the screen (3.4/5). Figure 3a shows a screenshot of the display during the treatment. A new function of the machine during the prescription phase, the "dose mode" (Fig. 2b), deserves a special mention. By setting the desired prescribed dose (e.g., 30 mL/kg/h [10]) and other parameters, like weight and hematocrit of the patient, the machine automatically calculates replacement (pre- or post-infusion) and/or dialysate flows. Furthermore, the cumulative dose value (Fig. 2a) that takes into account the downtime accumulated during the treatment, is continuously displayed and updated: this value can be compared with the prescribed dose and eventually leads to an adjustment of the clinical prescription, following the new concept of "dynamic prescription" proposed by 2016 ADQI group [11–14].

During alarm simulation phase, no critical issues were detected and an easy troubleshooting was experienced. Based on an internal risks analysis protocol, the time interval from the induction of the alarm to its detection by the machine was considered safe in all cases.

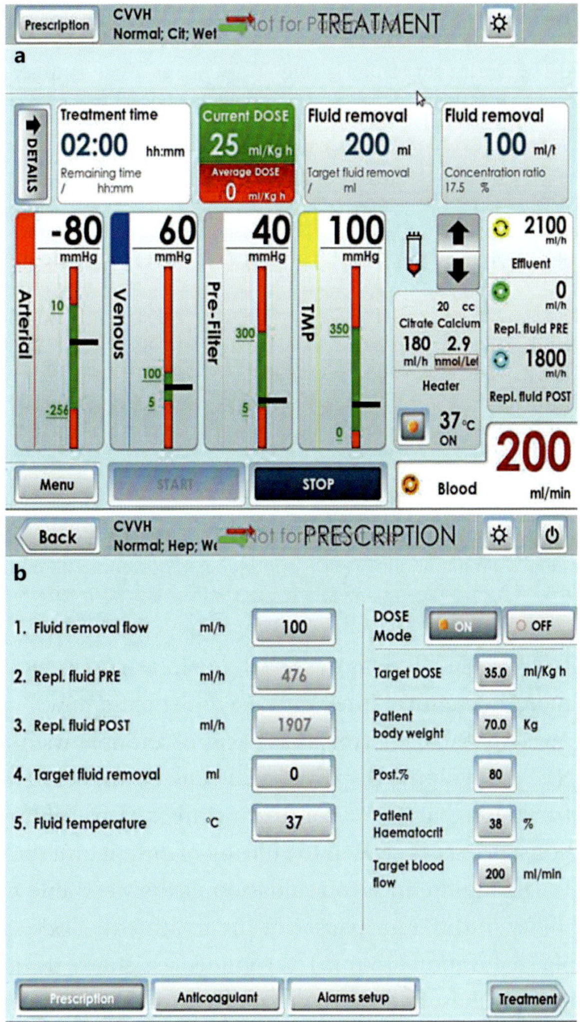

Fig. 3. a Screenshot of the treatment page. **b** Screenshot of the prescription page in "dose mode."

Finally, the precision of fluid balance system was analyzed. The absolute value of observed mass balance error at every time-sampling points varied less than 0.3% with respect to theoretical value of net ultrafiltration, as declared in the user manual of the machine.

In vivo Testing

Twelve CRRT daily treatments were performed in nine patients. Modalities and parameters are summarized in Table 2. Kibou® was utilized in CVVH (2), CVVHD (4), CVVHDF (3) and SCUF (3). On an average, a downtime of 5.30% of total treatment time (3.20% in CVVH, 5.39% in CVVHD, 9.70% in CVVHDF,

Table 2. Treatments and parameters performed with Kibou®

Treatment	Modality	Treatment time (min)	Blood flow (mL/min)	Pre replacement flow (mL/h)	Post replacement flow (mL/h)	Dialysate flow (mL/h)	NET UF flow (mL/h)	Anti-coagulation therapy	Anti-coagulation flow (UL/h)
1	CVVH	240	182	2,000	1,000	–	293	Heparin	667
2	CVVHD	240	200	–	–	3,000	400	NO	–
3	SCUF	190	200	–	–	–	600	Heparin	667
4	SCUF	240	220	–	–	–	700	NO	–
5	SCUF	165	190	–	–	–	400	NO	–
6	CVVHDF	270	200	–	2,000	2,000	500	Heparin	667
7	CVVHDF	180	190	–	1,000	7,000	10	Clexane	667
8	CVVH	180	200	2,000	1,000	–	600	NO	–
9	CVVHD	260	180	–	–	9,000	700	NO	–
10	CVVHDF	180	190	–	1,000	2,000	500	Heparin	667
11	CVVHD	240	225	–	–	7,000	200	NO	–
12	CVVHD	210	200	–	–	5,000	10	Heparin	667

NET UF, net ultrafiltration; CVVH, continuous veno-venous hemofiltration; CVVHD, continuous veno-venous hemodialysis; SCUF, slow continuous ultrafiltration; CVVHDF, continuous veno-venous hemodiafiltration.

3.22% in SCUF) was observed due to alarms detection, change bags procedure and other clinical interventions to the patient.

Fluid balance error at the end of the treatment, performed comparing the NET UF volume displayed real-time by the machine and values of bed scales, was lower than 0.3%, confirming the in vitro findings. During these CRRT sessions, no issues related to clotting of circuit and dialyzers were detected.

During the in vivo evaluation, users were able to better understand the usability of different phases of the treatment. They appreciated the low priming phase duration (13 min), the guided assistance of the machine during the preparation phase and the main screen during treatment, that was considered practical and immediate (3.7/5). Some difficulties were occasionally encountered during the disposable circuit placement on the machine (2.6/5).

Conclusions

A fourth generation machine for CRRT has been created and further developments are planned for the years to come. The new generation of machines is modifying the scenario of technology applied to critically ill patients. Some of these new technologies have not reached their full clinical potential yet. Nevertheless, they represent the future of this field. Our experience allows us to imagine how CRRT will evolve from today to the next ten years [15]. In addition to safety features of the therapy, other aspects have become important in further developments including smooth treatment performance, easy usability and consequent efficient use of resources, application of a standardized nomenclature to various parameters and possibly, a biofeedback-controlled treatment using signals from patients and machines to drive treatment parameters [16, 17].

Kibou® is a new multifunctional CRRT machine that meets most of the features required in a fourth generation CRRT equipment. Peculiarities and different aspects have been evaluated and described in this paper.

This represents a preliminary yet complete in vitro and short-time in vivo evaluation that should be soon completed by a comprehensive evaluation of the machine in a larger study. A larger-scale multicenter beta trial considering several CRRT sessions and modalities is scheduled to evaluate performance, safety and efficiency of the machine.

Acknowledgements

The authors would like to thank the clinical and nursing staff of Departments of Nephrology and Intensive Care Unit of S. Bortolo Hospital in Vicenza for their assistance and experience.

References

1 Ronco C: Evolution of technology for continuous renal replacement therapy: forty years of improvements. Contrib Nephrol 2017;189: 114–123.
2 Kramer P, Wigger W, Rieger J, Matthaei D, Scheler F: Arteriovenous haemofiltration: a new and simple method for treatment of over-hydrated patients resistant to diuretics]. Klin Wochenschr 1977;55:1121–1122.
3 Cruz D, Bobek I, Lentini P, Soni S, Chionh CY, Ronco C: Machines for continuous renal replacement therapy. Semin Dial 2009;22: 123–132.
4 Schlapfer P, Durovray JD, Plouhinec V, Chiappa C, Bellomo R, Schneider A: A first evaluation of OMNI®, a new device for continuous renal replacement therapy. Blood Purif 2017;43:11–17.
5 Ronco C, Garzotto F, Brendolan A, Zanella M, Bellettato M, Vedovato S, Chiarenza F, Ricci Z, Goldstein SL: Continuous renal replacement therapy in neonates and small infants: development and first-in-human use of a miniaturised machine (CARPEDIEM). Lancet 2014;383:1807–1813.
6 Ricci Z, Romagnoli S, Ronco C: Renal Replacement Therapy [Version 1; Referees: 4 Approved]. F1000 Research, 2016, p 5.
7 Clark WR, Garzotto F, Neri M, Lorenzin A, Zaccaria M, Ronco C: Data analytics for continuous renal replacement therapy: historical limitations and recent technology advances. Int J Artif Organs 2016;39:399–406.
8 Daugirdas JT, Ing TS: First-use reactions during hemodialysis: a definition of subtypes. Kidney Int Suppl 1988;24:S37–S43.
9 Baldwin I, Baldwin M, Fealy N, Neri M, Garzotto F, Kim JC, Giuliani A, Basso F, Nalesso F, Brendolan A, Ronco C: Con-current versus counter-current dialysate flow during CVVHD. A comparative study for creatinine and urea removal. Blood Purif 2016;41:171–176.
10 Ronco C, Bellomo R, Homel P, Brendolan A, Dan M, Piccinni P, La Greca G: Effects of different doses in continuous veno-venous haemofiltration on outcomes of acute renal failure: a prospective randomised trial. Lancet 2000;356:26–30.

11 Murugan R, Hoste E, Mehta RL, Samoni S, Ding X, Rosner MH, Kellum JA, Ronco C: Precision fluid management in continuous renal replacement therapy. Blood Purif 2016; 42:266–278.
12 Cerda J, Baldwin I, Honore PM, Villa G, Kellum JA, Ronco C: Role of technology for the management of AKI in critically ill patients: from adoptive technology to precision continuous renal replacement therapy. Blood Purif 2016;42:248–265.
13 Bagshaw SM, Chakravarthi MR, Ricci Z, Tolwani A, Neri M, De Rosa S, Kellum JA, Ronco C: Precision continuous renal replacement therapy and solute control. Blood Purif 2016; 42:238–247.
14 Ostermann M, Joannidis M, Pani A, Floris M, De Rosa S, Kellum JA, Ronco C: Patient selection and timing of continuous renal replacement therapy. Blood Purif 2016;42:224–237.
15 Clark WR, Neri M, Garzotto F, Ricci Z, Goldstein SL, Ding X, Xu J, Ronco C: The future of critical care: renal support in 2027. Crit Care 2017, in press.
16 Neri M, Villa G, Garzotto F, Bagshaw S, Bellomo R, Cerda J, Ferrari F, Guggia S, Joannidis M, Kellum J, Kim JC, Mehta RL, Ricci Z, Trevisani A, Marafon S, Clark WR, Vincent JL, Ronco C: Nomenclature for renal replacement therapy in acute kidney injury: basic principles. Crit Care 2016;20:318.
17 Villa G, Neri M, Bellomo R, Cerda J, De Gaudio AR, De Rosa S, Garzotto F, Honore PM, Kellum J, Lorenzin A, Payen D, Ricci Z, Samoni S, Vincent JL, Wendon J, Zaccaria M, Ronco C: Nomenclature for renal replacement therapy and blood purification techniques in critically ill patients: practical applications. Crit Care 2016;20:283.

Mauro Neri
International Renal Research Institute of Vicenza (IRRIV)
San Bortolo Hospital, Viale Rodolfi, 37
IT–36100 Vicenza (Italy)
E-Mail neri.mauro@gmail.com

La Manna G, Ronco C (eds): Current Perspectives in Kidney Diseases.
Contrib Nephrol. Basel, Karger, 2017, vol 190, pp 71–82 (DOI: 10.1159/000468915)

Phosphate in Chronic Kidney Disease Progression

Mario Cozzolino[a] · Denis Foque[b] · Paola Ciceri[a] · Andrea Galassi[a]

[a]Renal Division, Ospedale Santi Paolo e Carlo, Presidio San Paolo, Department of Health and Science, Università di Milano, Milan, Italy; [b]Department of Nephrology, University Lyon, UCBL, Carmen, Centre Hospitalier Lyon SUD, Pierre-Bénite, France

Abstract

A direct and independent association between serum phosphate (P) levels and mortality has been reported. High circulating P concentrations, still within the normal range, was associated with unfavourable outcomes in normal subjects as well as in chronic kidney disease (CKD) patients. Experimental data support the notion that P overload may hamper survival expectancy, directly inducing vascular, skeletal, and renal ageing. The balance of P results from dietary intake, intestinal absorption, glomerular filtration, tubular resorption, and hormonal asset. However, the accurate estimation of P balance is hampered by several methodological weaknesses, becoming even critical when renal function declines. Increasing evidence in the physiology of P metabolism in humans counteracts with uncertainties on dietary P intake and P balance assessment from general population and CKD subjects.

Introduction

Phosphate (P) is an essential element for life, being indispensable for energetic and mineral metabolism. However, both elevated and reduced serum P levels were associated with unfavourable outcomes. Clinical consequences of sustained hypophosphatemia may be relevant, as suggested by severe cellular and organ dysfunction observed in patients affected by inherited disorders of P metabolism [1]. On the contrary, hyperphosphatemia is accepted as a late consequence of advanced kidney disease, starting from glomerular filtration rate (GFR) lower than 30 mL/min [2–5]. High circulating P levels were repeatedly linked to reduced survival in observational studies, conducted among dialysis cohorts [6, 7]. A direct and independent association between P levels and adverse outcomes was even reported in normophosphatemic renal patients, including those with GFR greater than 30 mL/min [8–11], as well as in the general population [12–15]. Observational data showed that pre-dialysis and dialysis patients receiving P binders were exposed to lower mortality risk compared to controls [16–18]. Furthermore, high P load induced short life span in knock out animal models [19], which was counteracted by decreased dietary P intake [20, 21]. Experimental and epidemiological research shed light in the same direction, suggesting how increased P concentrations may be primarily responsible for poor clinical outcomes, by triggering and sustaining the Chronic Kidney Disease and Mineral Bone Disorder (CKD-MBD) syndrome, characterised by vascular ageing and altered mineral metabolism [2].

Although the best target of circulating P level to improve survival has never been tested in randomised controlled trials, the aforementioned data suggest that flexible thresholds of P concentration exist in human organism, beyond which P may elicit unsafe actions in addition to those considered indispensable for life. The committees of Kidney Disease Improving Global Outcomes [22] and the National Institute for Health and Care Excellence [23] have recently provided low graded suggestions to counteract the progressive increase of serum P levels with nutritional therapy and P binders, starting from GFR lower than 45 and 30 mL/min/1.73 m^2, respectively.

Regulation of P pool in humans depends on the homeostatic balance between nutritional intake, intestinal absorption and renal excretion [2]. Notably, the distribution of P concentrations is highly heterogeneous between body compartments and strictly regulated by hormonal assets. Both the net balance and the compartmental distribution of P are consequently expected to modulate the transition from homeostatic to maladaptive or pathologic actions of P. Consequently, nutritional intake, renal excretion and hormonal asset appear as fundamental modulators of the ratio between safe and unsafe effects elicited by P in humans. In this perspective, controlling dietary P intake may represent the easiest way to impact P balance through preventive and therapeutic interventions.

Two issues hamper the investigation of links between P intake and hard endpoints: (a) uncertainties on how to reliably assess the nutritional P intake and the net P balance in the long run, (b) disagreement on the most reliable marker of P load across different ranges of GFR. Thus, treating the link between P intake and mortality will require a cautious roadmap, starting with P balance in normal and renal patients, then moving to sources of nutritional P, mechanism of P toxicity and later dealing with observational and interventional studies on the topic.

P Homeostasis in Humans: Nutritional Intake, Intestinal Absorption and Renal Excretion

A 70 kg healthy man commonly stores 700 mg of P per day, mainly distributed as hydroxyapatite in skeletal tissues (85%) and to a lesser extent in intracellular space (14%) as organic P (nucleic acid, ATP, phospholipids, creatinine P and phosphoproteins). Circulating P represents only 1% of the total P amount, being consequently accepted as a weak marker of organic P pool and of its compartmental distribution [3].

Nutrients are the main source of P in humans. About 60% of dietary P is absorbed in the intestine as inorganic P, which can rise up to 80% in the presence of high circulating levels of calcitriol. Small intestine is deputed to P absorption by passive paracellular diffusion, regulated by electrochemical gradient, and by active transport elicited by luminal sodium P co-transporters type 2b (NPT2b) and type 2c (NPT2b) [24]. Once a certain amount of nutritional P is ingested, its intestinal uptake is mainly influenced by (a) vitamin D activity, (b) amount of P present in the diet, (c) bioavailability of P and (d) the presence of P binders [25]. At the same time serum P, circulating unbound to albumin, is continuously cleared by glomerular filtration, potentially leading to continuous P loss with unrelenting negative balance, if a net balance was not insured by tubular resorption of P (TRP). In normal subjects, under homeostatic conditions, about 75 and 10% of the filtered P are reabsorbed in proximal convoluted tubule, by 3Na-HPO4 co-transporter type 2a [26], and in distal convoluted tubule, respectively [3], with only 15% of the filtered P being excreted in urine. Thus, tubular handling is a leading guarantor of P equilibrium, actively responding to variations of nutritional intake, intestinal absorption and GFR under a strict hormonal regulation [26, 27]. An intestinal-glomerular feedback has also been postulated [28]. Based on this physiology consideration, the 24 h urinary P excretion (UPE) is generally considered equivalent to daily intestinal absorption (not intake) of P in normal subjects at a steady state [29, 30].

A positive P balance may result from the combination of one or more of the following factors: excessive nutritional intake, uncontrolled intestinal absorption and insufficient renal excretion. The reduction of GFR at levels higher than 60 mL/min may hamper the renal excretion of P, requiring the compensatory reduction of TRP, pivoted by fibroblast growth factor 23 (FGF23) [2, 4]. Assuming the maintenance of common nutritional intake and unchanged intestinal absorption of P, progressive reduction of GFR toward end-stage renal disease may become too pronounced for being counteracted by inhibited TRP, leading to inexorable positive P balance [2, 4, 5]. Phosphaturic response to a certain nutritional P load can be even subjected to individual predisposition [31–35], leading to heterogenous risk profile of positive P balance with relative clinical consequences. Dialysis patients are prone to severe risk of P overload, due to the minimal or null amount of renal excretion and the limited P clearance elicited by current dialysis techniques [2].

Although medical interventions are still far from optimizing the renal and dialysis clearance of P [2], nutritional P intake appears to be a more accessible target for preventive and therapeutic strategies. Any intervention focused on P intake should not be restricted to the whole amount of P delivered with meals, rather it should include the form of nutritional P (animal, vegetable or additive) and cooking methods, both capable of influencing nutritional P bioavailability [2]. In western countries, dietary P content ranges between 1,600 mg/day in adult men and 1,000 mg/day in adult women, despite the Recommended Dietary Allowance presents 700 mg as the suggested daily P intake [2, 12]. Animal proteins constitute the main source of dietary P, with a 60% bioavailability and an estimated average of 14 mg of P per g of proteins in a common mixed diet [36–38]. On the contrary, bioavailability of P of plant origin (as phytate) is less pronounced, due to the lack of phytase in humans (<50%), limiting the degradation of phytate to intestinal flora metabolism or to non-enzymatic hydrolysis reaction [39]. Main natural sources of P are consequently represented by cheese (220–700 mg/100 g), egg yolk (586 mg/100 g) and fish (190–290 mg/100 g) [39]. The choice of proteins with lower P-to-protein ratio is crucial, whenever reducing P intake is close to sufficient protein delivery. Another significant source of nutritional P is represented by additives as inorganic salts, which are almost entirely absorbed by intestinal tract (>90%) [39]. The widespread adoption of additives in soft drinks and low quality processed foods, close to poor labelling of P-containing preservatives, leads to considerable risk of "hidden" P load, especially in low socio-economic strata [2].

Cooking manipulation modulates the effective dietary P burden. Although boiling is considered a poor technique due to the loss of minerals, it is now a well-accepted method to reduce the content of P in animal and vegetable foods without a corresponding loss of proteins [39]. Boiling was capable of reducing the content of P up to 70% in flour, 51% in vegetables, 48% in legumes, 38% in meat and 19% in cheese [39].

Controlling P intake is a peculiar target of nutritional intervention in renal patients, commonly exposed to increased risk of P overload parallel to GFR reduction. Nutritional P load can be easily counteracted in pre-dialysis patients by reducing the protein intake to 0.8–0.6 g/kg/day, as suggested by current guidelines [22]. On the contrary, the higher recommended protein intake (1.2 g/kg/day) exposes dialysis patients to a considerable risk of nutritional P overload

[39], if not accompanied by careful selection of proteins with adequate P to protein ratio, avoidance of additive enriched foods and preference of adequate cooking methods [2].

Assessing Nutritional P Intake and P Balance: A Tough Goal in Research and Clinical Practice

A variety of methodological weaknesses hardly limits the assessment of nutritional P intake and P balance in clinical practice. An accurate estimation of P balance requires four steps: (a) assessment of P amount present in foods, (b) controlled amount of foods delivered, with consequent estimation of the P amount ingested with meals, (c) stools collection to compute the mass of nutritional P absorbed by intestinal tract, (d) assessment of P and creatinine renal excretion in 24 h urine collection (or in dialysate) to finally close the balance.

Although a similar approach is partially achievable only for research purposes in small and short intervention studies, it is difficult or even impossible to be perpetuated in large trials during the long follow-up, required to test the impact of nutritional therapy on hard endpoints. Thus, nutritional data collected from large cohorts are biased by restricted estimations of P intake, often limited to baseline prescribed P dose, dietary records validated on pre-coded menu books, dietary questionnaire, 24-h dietary recall and 1-month food frequency questionnaire.

Assessing 24-h UPE is considered an alternative and reliable estimate of intestinal P absorption [29, 30]. However, its reliability is limited to the period of urine collection, generally confined to the study baseline and to further control visits, rather than representing the averaged nutritional P load during the entire length of follow-up. Furthermore, assuming that a direct correspondence between 24-h UPE and intestinal absorption of P is also maintained in renal cohorts, especially at GFR lower than 30 mL/min, would imply the jarring assumption of a neutral P balance in advanced chronic kidney disease (CKD), which is contradictory with the common observation of increased serum P levels in CKD stages 4–5 [2–4]. Data from the Modification of Diet in Renal Disease clearly showed that patients with lower GFR (13–24 mL/min/1.73 m^2) experienced a progressive increase of serum P levels compared to patients with higher GFR (25–55 mL/min/1.73 m^2) despite the lower UPE, consequent to reduced P intake with low and very low protein diet (350–700 and 280–630 mg/day of P, respectively) [39]. Authors randomised 18 CKD patients (GFR 15–60 mL/min) and 12 healthy controls to 3 interventions: (a) high P diet (2,000 mg/day), (b) low P diet (750 mg/day), (c) low P diet plus aluminium hydroxide 500 mg thrice daily at meals [39]. Significant interaction between CKD status and the effect of P intake on absolute change of UPE and FGF23 was detected. In this context, Phelps et al. [33, 34] have recently observed that, at the steady state of reduced GFR, serum P is determined by both the influx into extracellular space and TRP. Notably, the conclusion by Phelps et al. [33, 34] were derived by fasting specimens of blood and urine in the morning, assuming that P homeostasis should be evaluated during the brief period required to create an aliquot of urine, although not being representative of the continuous variations that occurred during 24 h. However, it can be argued that tubular handling of P may elicit a deep impact on P balance at reduced GFR, limiting the reliability of 24-h UPE alone as an exhaustive marker of intestinal P absorption. Furthermore, observational data proposed a strong impact on mortality elicited by tubular sensitivity to the phosphaturic action of FGF23, represented by FGF23/TRP ratio [32].

Caution should be also taken while dealing with serum P as a marker of dietary P intake and P balance. Among 15,513 patients, enrolled in the Third National Health and Nutrition Examination Survey (NHANES III), dietary P intake assessed by 24-h dietary recall was poorly associated with serum P (0.03 mg/dL higher serum P for each 500 mg greater intake of P) only in subjects with GFR greater than 60 mL/min. A weak association between dietary P and circulating P levels was observed among 1.105 CKD patients from the same cohort, with a 0.009 mg/dL increase of serum P for each 100 mg increase of daily P intake. A non-significant 0.03 mg/dL higher serum P concentration was similarly reported for each 300 mg/day greater UPE in the Heart and Soul Study ($p = 0.07$) [30]. Furthermore, a single sample poorly represents the circadian fluctuations of serum P levels, characterised by a nadir around 11:00 a.m. and to further peaks at 04:00 p.m. and at 01:00–03:00 a.m. [39]. In the NHANES III study, serum P levels were considerably higher when assessed in the afternoon, in the evening and at non-fasting status, compared to the morning and fasting sessions [39].

Concerns even exist about the entity of P intake variations, required to induce detectable modification of mean circulating P levels within a short period of time (Table 1). Reduction of P intake from 1,500 mg to less than 500 mg was associated with 40% reduction of circulating P levels and descriptive mitigation of diurnal picks in healthy men, detectable after 1 day and even more pronounced after 10 days of diet modification. Notably, serum P levels assessed at 08:00 were unchanged after the first 24 h of diet restriction. Doubling P intake to 3,000 mg for 10 days led to a 14% increase of the 24 h mean P levels without significant variations of the fasting morning levels and with an exaggerated early afternoon pick. Increased serum P levels were similarly observed after switching P intake from 850 to 2,880 mg for 36 h. On the con-

Table 1. Principal studies investigating variations of serum P levels after controlled interventions on dietary P intake

Study	Subjects	Kidney function	Diet intervention on P	Follow-up	Main results
Portale et al. [40], 1987	6 healthy men	–	1,500 mg/day/70 kg for 9 days; 500 mg/day plus aluminum hydroxide for 10 days; 3,000 mg/day for 10 days	29 days	40% reduction of serum P levels after 10 days of P restriction; 14% increase in 24 h mean serum P after 10 days of P supplementation
Vervloet et al. [41], 2011	18 healthy subjects	CrCl 125 + 27 mL/min	850 mg/day for 3 days; 1 week interval; 2,880 mg/day for 3 days	13 days	No change in serum P after P restriction; increased serum P from averaged 1.10 + 0.09 mmol/L to a maximum of 1.32 mmol/L after high dietary intake
Kremsdof [42], 2013	19 healthy volunteers	eGFR >60 mL/ min/1.73 m^2	1,156 + 271 mg/day for 13 days; 2,071 + 346 mg/day for 14 days; 1,622 + 442 mg/day for 101–115 days	14 weeks	No change in serum P levels observed after diet modification

Studies are presented as principal Author's name and year of publication. eGFR, estimated glomerular filtration rate; P, phosphate.

trary, a milder increase of nutritional P intake (from 1,556 to 2,071 mg/day) did not induce any detectable change in mean serum P levels, assessed by serial blood draws for 24 h after 2 weeks of diet modification in 19 healthy volunteers [39].

Circulating P levels may also be modulated by gender, age and race [35]. However, the observational source of data does not allow any conclusion about an active influence elicited by age, gender and race on P balance and, even more, about their interaction with the reliability of serum P levels as a marker of P balance within specific patient categories.

Serum P was even insensible to fixed doses of P binders and diet intervention in pre-dialysis CKD patients on the short run, despite a significant reduction of UPE. Isakova et al. randomly assigned 16 normophosphatemic CKD 3–4 patients to controlled diet intervention for 2 weeks (1,500 vs. 750 mg of P) associated with lanthanum carbonate 1 g thrice daily versus placebo [17]. Of note, circulating P and FGF23 levels were unchanged despite significant reduction of 24-h UPE in patients on low P diet (249 ± 213 vs. 702 ± 262 mg) and lanthanum carbonate (from 710 ± 192 mg at baseline to 267 ± 140 mg at the end of follow-up). Thirty-nine normophosphatemic CKD patients (mean estimated GFR 37.8 ± 11.0 mL/min) were randomised to nutritional intervention (ad libitum diet versus 900 mg of P intake), alternatively associated with placebo versus lanthanum carbonate 1,000 mg thrice daily for 3 months. Difference in daily P intake, estimated by 3-day food records, was lower than 500 mg between the study arms. Although 3 g of lanthanum is expected to reduce intestinal P absorption by 405 mg, the difference in 24-h UPE between patients receiving controlled diet and active lanthanum was lower than expected, compared to those allocated to placebo plus similar diet intervention (544 ± 294 vs. 757 ± 371 at study end) [38]. Notably, serum P levels and TRP were insensitive to study interventions, and were independent of the widely ranged reduction of FGF23 levels (35 ± 32%). Eighteen normophosphatemic CKD stage 3 patients received controlled P intake lower than 700 mg for 28 days plus lanthanum carbonate 750 mg thrice weekly for 4 additional weeks). UPE was assessed in the 12-h urine sampled at fasting night time. Consistent with previous results, serum P concentration did not vary despite significant reduction of UPE and FGF23 levels. Authors randomised 40 normophosphatemic CKD stages 3–4 patients to calcium carbonate versus sevelamer carbonate at escalating doses for 6 weeks (up to 5.28 and 6.4 g of calcium carbonate and sevelamer, respectively). Patients were instructed to maintain a constant P intake of approximately 600 mg/day. P levels remained unchanged, although they were accompanied by reduction of 24-h UPE, FGF23 levels and fractional excretion of P. Despite the aforementioned results, repeated evidences showed that significant reduction of serum P levels can be achieved in CKD patients by administering P binders at increasing doses, properly titrated to reduce serum P levels toward a specific target [37].

In conclusion, assessing P load remains challenging in clinical practice. Hypo and hyperphosphatemia are well accepted markers of contracted and expanded P pool, respectively. Variations of P intake greater than 1,000 mg/day repeatedly induced acute modification of serum P levels in normal subjects. On the contrary, phosphatemia was insensitive to lower modification of nutritional P intake in pre-dialysis CKD patients within few weeks. Thus, how to early estimate the small changes in P balance remains an unsolved issue among normophosphatemic CKD patients. It can be argued that small modification of nutritional intake, intestinal absorption and renal handling can impact the P pool and circulating P levels in the long run, with a significant interaction with both the GFR and the tubular sensitivity to hormonal adaptive response. A hot topic in the field is currently represented by the research of a coveted surrogate marker, sufficiently reliable to provide an accurate signal of P overload and of its toxic activity in still normophosphatemic subjects. FGF23 and TRP have been proposed in this regard, although with still inconsistent results [39].

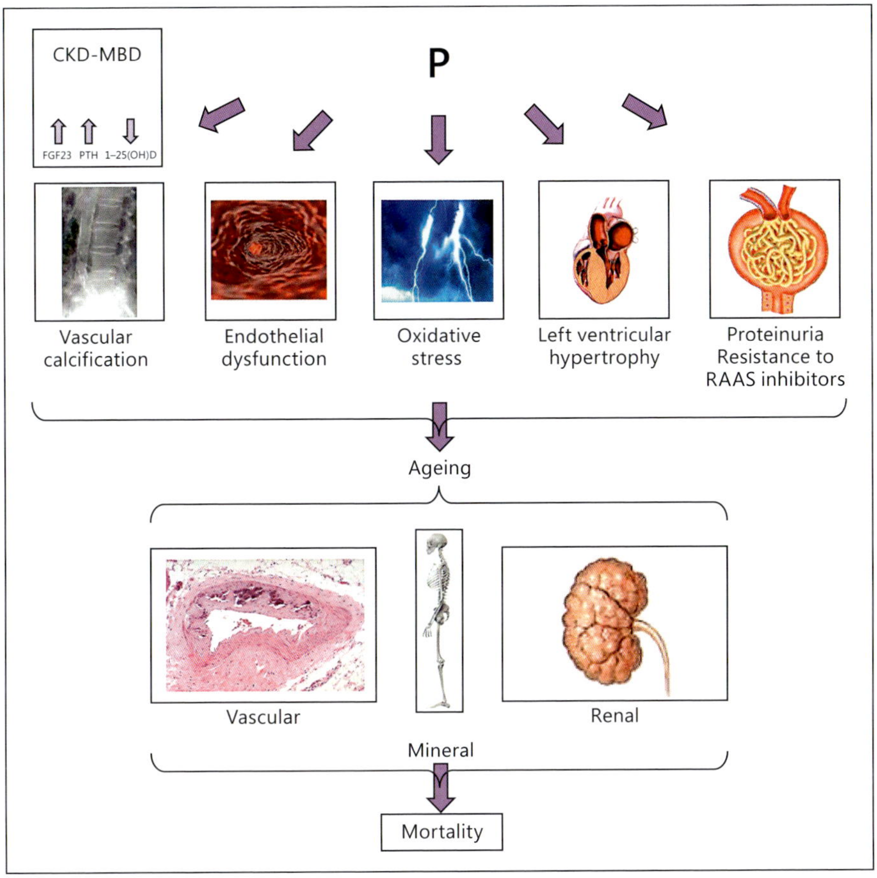

Fig. 1. Mechanisms linking P to mortality. CKD-MBD, chronic kidney disease and mineral bone disorder; FGF23, fibroblast growth factor 23; P, phosphate; PTH, parathyroid hormone; RAAS, renin-angiotensin-aldosterone-system.

P and Mortality: Mechanisms in Pills

Detailed mechanisms linking P to poor outcomes will be examined in chapters focused on P intake and cardiac-bone disease. It is herein sufficient to synthesise that P overload may negatively impact the survival accelerating vascular, skeletal and renal ageing through the following pathways (Fig. 1): (a) activation of CKD-MBD syndrome pivoted by FGF23 [2], (b) endothelial dysfunction and vascular calcification [2], (c) oxidative stress, (d) left ventricular hypertrophy sustained by high FGF23 levels close to the downregulation of its co-receptor Klotho, and (e) glomerular damage with worsening of proteinuria and GFR, associated with an impaired response to renin-angiotensin-aldosterone system inhibitors.

Conclusions

The impact of P load on survival remains a hot topic for contemporary medicine. Improving the knowledge on P balance and its potential toxicity is challenging to improve the care of general population and renal patients. Although both hypo and hyperphosphatemia have been related to poor outcomes, hard uncertainties persist about the negative impact elicited by incremental P balance on normophosphatemic individuals in general population and, especially, in pre-dialysis CKD cohorts. Further insights are urgently needed to bridge the gaps still existing within and between the notions of pathophysiology and epidemiology in this regard. Even more, holistic approach to nutrition should never be abandoned, always considering P as a crucial, but not unique, element of the diet close to calories, proteins carbohydrates and other elements. Awaiting dedicated randomised controlled trials on the topic, disposable evidences are mainly observational and still insufficient to recommend any different intervention from those purposed by the current guidelines.

References

1 Wagner CA, Rubio-Aliaga I, Biber J, Hernando N: Genetic diseases of renal phosphate handling. Nephrol Dial Transplant 2014; 29(suppl 4):iv45–iv54.
2 Galassi A, Cupisti A, Santoro A, et al: Phosphate balance in ESRD: diet, dialysis and binders against the low evident masked pool. J Nephrol 2015;28:415–429.
3 Uribarri J: Phosphorus homeostasis in normal health and in chronic kidney disease patients with special emphasis on dietary phosphorus intake. Semin Dial 2007;20:295–301.
4 Isakova T, Wahl P, Vargas GS, et al: Fibrobalst growth factor 23 is elevated before parathyroid hormone and phosphate in chronic kidney disease. Kidney Int 2011;79:1370–1378.
5 Bellasi A: Pro: should phosphate binders be used in chronic kidney disease stage 3–4? Nephrol Dial Transplant 2016;31:184–188.
6 Block GA, Klassen PS, Lazarus JM, et al: Mineral metabolism, mortality, and morbidity in maintenance hemodialysis. J Am Soc Nephrol 2004;15:2208–2218.
7 Kalantar-Zadeh K, Kuwae N, Regidor DL, et al: Survival predictability of time-varying indicators of bone disease in maintenance hemodialysis patients. Kidney Int 2006;70:771–780.
8 Bellasi A, Mandreoli M, Baldrati L, et al: Chronic kidney disease progression and outcome according to serum phosphorus in mild-to-moderate kidney dysfunction. Clin J Am Soc Nephrol 2011;6:883–891.
9 Kestenbaum B, Sampson JN, Rudser KD, et al: Serum phosphate levels and mortality risk among people with chronic kidney disease. J Am Soc Nephrol 2005;16:520–528.
10 Menon V, Greene T, Pereira AA, et al: Relationship of phosphorus and calcium-phosphorus product with mortality in CKD. Am J Kidney Dis 2005;46:455–463.
11 Kovesdy CP, Ahmadzadeh S, Anderson JE, Kalantar-Zadeh K: Secondary hyperparathyroidism is associated with higher mortality in men with moderate to severe chronic kidney disease. Kidney Int 2008;73:1296–1302.

12 Dhingra R, Sullivan LM, Fox CS, et al: Relations of serum phosphorus and calcium levels to the incidence of cardiovascular disease in the community. Arch Intern Med 2007;167: 879–885.
13 Onufrak SJ, Bellasi A, Shaw LJ, et al: Phosphorus levels are associated with subclinical atherosclerosis in the general population. Atherosclerosis 2008;199:424–431.
14 Foley RN, Collins AJ, Ishani A, et al: Calcium-phosphate levels and cardiovascular disease in community-dwelling adults: the Atherosclerosis Risk in Communities (ARIC) Study. Am Heart J 2008;156:556–563.
15 Linefsky JP, O'Brien KD, Katz R, et al: Association of serum phosphate levels with aortic valve sclerosis and annular calcification: the cardiovascular health study. J Am Coll Cardiol 2011;58:291–297.
16 Kovesdy CP, Kuchmak O, Lu JL, et al: Outcomes associated with phosphorus binders in men with non-dialysis-dependent CKD. Am J Kidney Dis 2010;56:842–851.
17 Isakova T, Gutierrez OM, Chang Y, et al: Phosphorus binders and survival on hemodialysis. J Am Soc Npehrol 2009;20:388–396.
18 Cannata-Andia JB, Fernandez-Martin JL, Locatelli F, et al: Use of phosphate-binding agents is associated with a lower risk of mortality. Kidney Int 2013;84:998–1008.
19 Kuro-o M: Klotho and aging. Biochim Biophys Acta 2009;1790:1049–1058.
20 Razzaque MS, Sitara D, Taguchi T, St-Arnaud R, Lanske B: Premature aging-like phenotype in fibroblast growth factor 23 null mice is a vitamin D-mediated process. FASEB J 2006;20:720–722.
21 Stubbs JR, Liu S, Tang W, et al: Role of hyperphosphatemia and 1,25-dihydroxyvitamin D in vascular calcification and mortality in fibroblastic growth factor 23 null mice. J Am Soc Nephrol 2007;18:2116–2124.
22 Kidney Disease: Improving Global Outcomes (KDIGO) CKD Work Group: KDIGO 2012 clinical practice guideline for the evaluation and management of chronic kidney disease. Kidney Int Suppl 2013;3:1–150.
23 National Institute for Health and Care Excellence (NICE): Hyperphosphatemia in Chronic Kidney Disease. Disponibile su guidance. nice.orguk/cg157, 2013.
24 Allen LH, Wood RJ: Calcium and phosphorus in modern nutrition; in Shils ME, Olson JA, Shike M (eds): Health and disease, ed 8th. Lea & Febiger, Philadelphia, 1994, pp 144–163.
25 Ramirez JA, Emmett M, White MG, et al: The absorption of dietary phosphorus and calcium in hemodialysis patients. Kidney Int 1986;30:753–759.
26 Tenenhouse HS:Regulation of phosphate homeostasis by the type IIaNa/phosphate cotransporter. Ann Rev Nutr 2005;25:197–214.
27 Berndt TJ, Schiavi S, Kumar R, et al: "Phosphatonins" and the regulation of phosphorus homeostasis. Am J Physiol Renal Physiol 2005;189:F1170–F1182.
28 Marks J, Debnam ES, Unwin RJ: Phosphate homeostasis and the renal-gastrointestinal axis. Am J Physiol Renal Physiol 2010; 299:F285–F296.
29 Selamet U, Tighiouart H, Sarnak MJ, et al: Relationship of dietary phosphate intake with risk of end-stage renal disease and mortality in chronic kidney disease stages 3–5: the Modification of Diet in Renal Disease Study. Kidney Int 2016;89;176–184.
30 Palomino HL, Rifkin DE, Anderson C, et al: 24-hour urine phosphorus excretion and mortality and cardiovascular events. Clin J Am Soc Nephrol 2013;8:1202–1210.
31 Yoda K, Imanishi Y, Yoda M, et al: Impaired response of FGF-23 to oral phosphate in patients with type 2 diabetes: a possible mechanism of atherosclerosis. J Clin Endocrinol Metab 2012;97:E2036–E2043.
32 Dominguez JR, Shlipak MG, Whooley MA, et al: Fractional excretion of phosphorus modifies the association between fibroblast growth factor-23 and outcomes. J Am Soc Nephrol 2013;24:647–654.
33 Phelps KR, Mason DL: Parameters of phosphorus homeostasis at normal and reduced GFR: theoretical considerations. Clin Nephrol 2015;83:167–176.
34 Phelps KR, Mason DL, Stote KS: Parameters of phosphorus homeostasis at normal and reduced GFR: empiric observations. Clin Nephrol 2015;83:208–217.
35 Cirillo M, Ciacci C, DE Santo NG: Age, renal tubular phosphate reabsorption, and serum phosphate levels in adults. N Engl J Med 2008;359:864–866.

36 Boaz M, Smetana S: Regression equation predicts dietary phosphorus intake from estimate of dietary protein intake. J Am Diet Assoc 1996;96:1268–1270.
37 Noori N, Kalantar-Zadeh K, Kovesdy CP, et al: Association of dietary phosphorus intake and phosphorus to protein ratio with mortality in hemodialysis patients. Clin J Am Soc Nephrol 2010;5:683–692.
38 D'Alessandro C, Piccoli GB, Cupisti A: The "phosphorus pyramid": a visual tool for dietary phosphate management in dialysis and CKD patients. BMC Nephrol 2015;16:9.
39 Cozzolino M, Bruschetta E, Cusi D, et al: Phosphate handling in CKD-MBD from stage 3 to dialysis and the three strengths of lanthanum carbonate. Expert Opin Pharmacother 2012;13:2337–2353.
40 Portale AA, Halloran BP, Morris RC Jr: Dietary intake of phosphorus modulates the circadian rhythm in serum concentration of phosphorus. Implications for the renal production of 1,25-dihydroxyvitamin D. J Clin Invest 1987;80:1147–1154.
41 Vervloet MG, van Ittersum FJ, Büttler RM, Heijboer AC, Blankenstein MA, ter Wee PM: Effects of dietary phosphate and calcium intake on fibroblast growth factor-23. Clin J Am Soc Nephrol 2011;6:383–389.
42 Kremsdorf RA, Hoofnagle AN, Kratz M, Weigle DS, Callahan HS, Purnell JQ, Horgan AM, de Boer IH, Kestenbaum BR: Effects of a high-protein diet on regulation of phosphorus homeostasis. J Clin Endocrinol Metab 2013;98:1207–1213.

Mario Cozzolino, MD, PhD, FERA, FASN
Renal Division, San Paolo Hospital
Department of Health Sciences, University of Milan
Via A. di Rudinì, 8, IT–20142 Milano (Italy)
E-Mail mario.cozzolino@unimi.it

La Manna G, Ronco C (eds): Current Perspectives in Kidney Diseases.
Contrib Nephrol. Basel, Karger, 2017, vol 190, pp 83–95 (DOI: 10.1159/000468952)

Fibroblast Growth Factor 23: Mineral Metabolism and Beyond

Alexander Grabner[a] · Sandro Mazzaferro[b] · Giuseppe Cianciolo[c] · Stefanie Krick[d] · Irene Capelli[c] · Silverio Rotondi[b] · Claudio Ronco[e, f] · Gaetano La Manna[c] · Christian Faul[g]

[a]Division of Nephrology, Department of Medicine, Duke University Medical Center, Duke University, Durham, NC, USA; [b]Department of Cardiovascular, Respiratory, Nephrologic, Anesthetic and Geriatric Sciences, Sapienza University of Rome, Rome, and [c]Department of Experimental Diagnostic and Specialty Medicine (DIMES); Nephrology, Dialysis and Renal Transplant Unit, St. Orsola Hospital, University of Bologna, Bologna, Italy; [d]Division of Pulmonary, Allergy, Critical Care and Sleep Medicine, Department of Medicine, University of Alabama at Birmingham, Birmingham, AL, USA ; [e]Department of Nephrology, Dialysis and Transplantation, and [f]International Renal Research Institute of Vicenza, San Bortolo Hospital, Vicenza, Italy; [g]Katz Family Drug Discovery Center and Division of Nephrology and Hypertension, Department of Medicine, University of Miami Leonard M. Miller School of Medicine, Miami, FL, USA

Abstract

Patients affected by chronic kidney disease (CKD) exhibit a high risk of cardiovascular mortality that is poorly explained by traditional risk factors. There is a growing awareness about the role of derangement of mineral metabolism that is currently accepted as a trigger and sustainer of cardiovascular disease (CVD) in CKD patients. The synthetic definition of CKD mineral and bone disorder (CKD-MBD) split the concept that the indexes of mineral metabolism extend their effects beyond the bone until the vascular wall and metabolic milieu of CKD patients through complex pathways. A better understanding of the biomarkers and mechanisms of left ventricular hypertrophy, CVD, inflammation, and chronic renal damage may help with the diagnosis and treatment of the systemic impairment that occurs secondary to CKD-MBD, thus slowing the progression of renal and CVD and improving patient survival. Recent insights into fibroblast growth factor (FGF) 23 have led to marked advancement in interpreting data on CVD and CKD progression ascribing to FGF23 a pivotal role in these pathologies independent of its co-receptor klotho and well beyond mineral metabolism. This review article will discuss the current experimental and clinical evidence regarding the role of FGF23 in physiology and pathophysiology of CKD and its associated complications with an emphasis on CVD.

A.G., S.M., and G.C. contributed equally to this work.

Physiology and Pathophysiology of Fibroblast Growth Factor 23

Primarily secreted by osteocytes, fibroblast growth factor (FGF) 23 is a hormone mainly involved in the regulation of mineral metabolism. In the kidney and the parathyroid glands, FGF23 binds FGF receptor (FGFR)/klotho co-receptor complexes to reduce serum phosphate levels (Fig. 1), inhibit parathyroid hormone (PTH) secretion, and decrease levels of active vitamin D [1]. Specifically in the kidney, FGF23 induces urinary phosphate excretion by decreasing the expression of sodium-phosphate co-transporters in the proximal tubule apparatus [2]. A reduction in active vitamin D levels is achieved by inhibition of 1-α-hydroxylase, which catalyzes the hydroxylation of 25-hydroxyvitamin D to 1,25-dihydroxyvitamin D_3 and by stimulation of 24-hydroxylase, which converts 1,25-dihydroxyvitamin D_3 to inactive metabolites in the proximal tubule [3]. In the distal tubule, FGF23 has been shown to augment calcium and sodium reabsorption through increased apical expression of epithelial calcium channel TRPV5 and the sodium-chloride co-transporter [4]. Furthermore, FGF23 suppresses the expression of angiotensin converting enzyme-2 in the kidney, thereby leading to an activation of the renin-angiotensin-aldosterone-system (RAAS). Phosphate load, 1,25-dihydroxyvitamin D_3, and PTH belong to the main group of physiologic regulators of FGF23 synthesis. However, several additional factors including calcium, the RAAS, oxidative stress, parameters of iron metabolism, and inflammation have been shown to regulate FGF23 production and secretion from osteocytes [5]. Nevertheless, the complete mechanisms behind the production and secretion of FGF23 from osteocytes remain poorly understood, are definitely complex, and also involve a number of local factors such as dentin matrix protein 1 or phosphate regulating endopeptidase homolog X-linked.

In patients with chronic kidney disease (CKD), serum levels of FGF23 rise progressively as kidney function declines. This response is mainly a compensatory mechanism to maintain neutral phosphate balance by promoting additional urinary phosphate elimination to counteract the defect in renal excretory capacity. Several large epidemiological studies demonstrated a powerful dose-dependent association between serum levels of FGF23 and higher risk of mortality in end-stage renal disease (ESRD) patients. Moreover, higher FGF23 correlates with increased prevalence of cardiovascular disease (CVD) in general and left ventricular hypertrophy (LVH) in particular among CKD patients. Several in vitro and in vivo studies have been performed to identify a potential causative role of FGF23 in the pathophysiology of abnormal cardiac remodeling in CKD, also known as uremic cardiomyopathy. FGF23 induces hypertrophic growth of cardiac myocytes in vitro. Moreover, rodent models with elevated serum FGF23

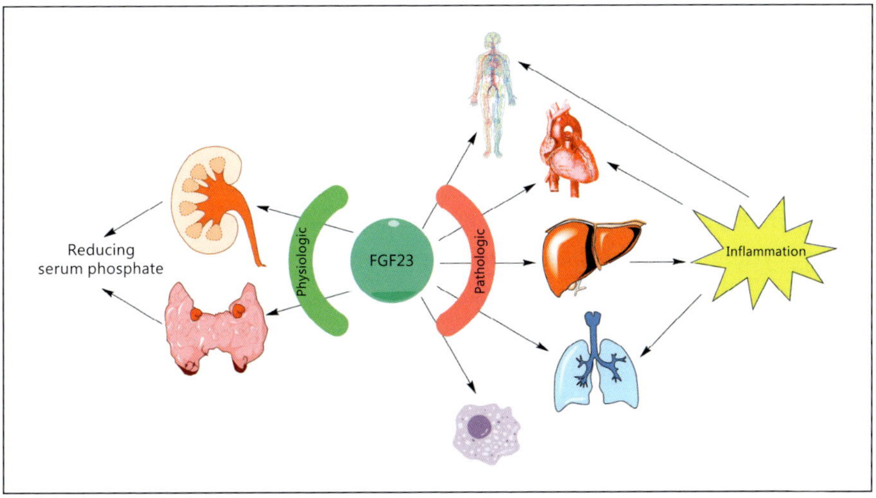

Fig. 1. Renal and extrarenal effects of FGF23. In a physiologic state, FGF23 mainly targets the kidney and the parathyroid glands to maintain phosphate homeostasis. In CKD, elevated FGF23 levels might contribute to endothelial dysfunction, cause left ventricular hypertrophy, and promote a chronic inflammatory state. Moreover, FGF23 interferes with the immune system by impairing neutrophil granulocytes. Finally, FGF23 might also account for systemic inflammation observed in COPD. Prolonged chronic inflammation then further accelerates cardiovascular disease.

levels, either by injection of recombinant FGF23, application of a high phosphate diet or induction of CKD using surgical renal ablation, develop cardiac hypertrophy [6]. Detailed analyses of FGF23-induced signaling events have identified cardiac FGFR4 as the isoform mediating klotho-independent effects of FGF23 on the heart [7]. Subsequent downstream signaling involves the stimulation of the phospholipase Cγ (PLCγ)/calcineurin/nuclear factor of activated T cells (NFAT) pathway, which acts as a potent inducer of cardiac hypertrophy in response to other pathological stimuli. Global deletion of FGFR4 protects mice from the development of high phosphate diet-mediated LVH, whereas systemic blockade of FGFR4 with an isoform-specific blocking antibody or inhibition of calcineurin prevents the development of cardiac hypertrophy in the 5/6 nephrectomy rat model of CKD. Noteworthy, these treatment studies in rats did not affect hypertension or ameliorate kidney function, indicating that high blood pressure or other factors altered by reduced kidney function do not play a major role in the development of LVH at least in this particular animal model of CKD. Finally, knockin mice bearing a gain-of-function mutation of FGFR4, characterized by a single amino acid substitution at position 385 (glycine to arginine), spontaneously develop LVH at 6 months of age [7]. Taken together, these data provide strong evidence of the deleterious effects of FGF23-mediated activation of cardiac FGFR4. Nevertheless, further in vivo studies with cell-type specific deletion of FGFR4 in the presence of elevated FGF23 are needed to confirm whether FGFR4 in cardiac myocytes is required for the development of LVH.

Clinical studies have not only associated FGF23 with cardiovascular injury and LVH, but also with inflammation and elevated serum levels of pro-inflammatory cytokines (Fig. 1). Since the liver expresses high levels of FGFR4, lacks klotho, and plays a decisive role in the acute-phase response and the expression of pro-inflammatory markers, it is intriguing to speculate that FGF23 directly contributes to inflammation in CKD. In a variety of in vitro experiments, Singh et al. [8] demonstrated that FGF23 also activates FGFR4 and subsequent PLCγ/calcineurin/NFAT signaling in hepatocytes, which leads to the production of pro-inflammatory cytokines such as interleukin-6 (IL-6) and C-reactive protein (CRP). Elevated serum FGF23 levels increase hepatic and circulating levels of IL-6 and CRP in wild-type but not in FGFR4 knockout mice. Furthermore, in 5/6 nephrectomized rats, the administration of a FGFR4-blocking antibody reduces the hepatic expression and circulating levels of CRP and IL-6. These findings indicate a novel mechanism for chronic inflammation in CKD and suggest that FGFR4 blockade might not only ameliorate the cardiovascular burden of CKD, but also have therapeutic anti-inflammatory effects.

CKD is also associated with impaired host response and increased susceptibility to infections. Recently, FGF23 has been identified to inhibit chemokine-activated leukocyte arrest on the endothelium. Antibody-mediated neutralization of FGF23 restored leukocyte recruitment and host defense in 5/6 nephrectomized mice with pneumonia, thereby significantly improving the survival in this particular animal model [9]. Mechanistically, FGF23 binds to FGFR2, activates protein kinase A, and inhibits the small GTPase Rap1, thereby counteracting selectin and chemokine-triggered activation of β_2-integrin in neutrophil granulocytes [9]. Nevertheless, further mechanistic studies are needed to elucidate the therapeutic potential of targeting FGF23 for the control of inflammatory disorders, especially in the context of CKD (Fig. 1).

Systemic inflammation is not only associated with CKD, but also with a hallmark of other diseases such as chronic obstructive pulmonary disease (COPD), which is associated with increased oxidative stress and significantly increased circulating pro-inflammatory cytokines. Therefore, the question was raised whether FGF23 could serve as a valid biomarker to indicate the aging process in COPD. To date, only a few reports have shown that serum FGF23 levels correlate with smoking [10], and direct effects of FGF23 on the lung have not been described. However, it has been demonstrated that klotho is reduced in COPD patients and protects the alveolar epithelium against oxidative damage [11]. Furthermore, klotho-deficient mice develop widened alveolar spaces, consistent with pulmonary emphysema in the presence of high plasma FGF23 levels [8]. Certainly, future translational studies will assess further potential pathophysiologic roles of FGF23 in a complex scenario such as CKD or COPD.

FGF23 and Clinical Outcomes in CKD and ESRD

In CKD patients, the first striking clinical observation of the potential pathophysiologic role of FGF23 came from a study on incident dialysis patients showing that increased levels of FGF23 associated with increased mortality, independently of the most accredited culprit, serum levels of phosphate. In this study, involving 10,044 patients who were followed up for 1 year, the authors found an impressive monotonic increase in the rate of death, along with increasing FGF23 quartiles, in a nested case–control sample of patients (200 cases per group) [12]. Notably, in these patients, serum levels of FGF23 were up to a thousand times higher than normal. Similar results were evident in the CRIC study involving 3,879 chronic renal failure patients not on dialysis, who were followed for a median of 3.5 years [13]. In this large cohort of patients, higher quartiles of FGF23 were associated with an increased risk of mortality and also of entering renal replacement therapy. By comparison, in this population the median value of FGF23 was only 3-fold higher than normal. Finally, in renal transplantation, higher tertiles of FGF23 associated with increased mortality and worse renal outcome, even though median serum levels of FGF23 were within the normal range [14]. Conceivably, the apparent toxicity associated with serum FGF23 seems to occur for any degree of increment, a finding that raises the question of the underlying pathologic mechanism. Indeed, further data from the CRIC study showed that higher values of FGF23 associated with echocardiographic evidence of LVH, which suggested a possible cardiotoxicity [6].

Interestingly, a recent study evaluated the presence of FGF23 not in bone but in myocardial autopsy samples of deceased chronic renal failure patients. Compared to subjects with normal renal function, the expression of FGF23, FGFR4, and of the calcineurin pathway (involved with myocardial remodeling and hypertrophy) was significantly increased in dialysis patients [15]. Although obtained in a limited number of cases, these results clearly support the link between FGF23 and myocardial uremic toxicity. Furthermore, observational data in conservative chronic renal failure (CRF) analyzing the correlation of serum phosphate and of FGF23 with vascular or cardiac events, clearly suggested that hyperphosphatemia mainly contributes to vascular calcification while FGF23 mainly contributes to myocardial hypertrophy [16].

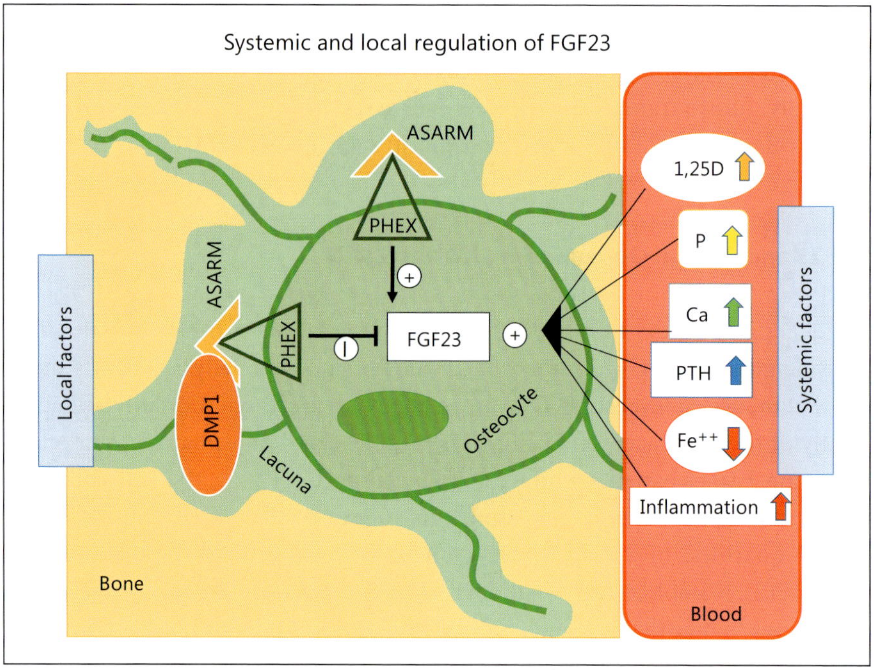

Fig. 2. Systemic and local factors regulating osteocyte production of FGF23. Local: DMP1-ASARM binds PHEX and suppresses FGF23 expression; free ASARM-peptide binds PHEX preventing the binding to DMP1 and increased expression of FGF23. Systemic: 1,25D, phosphate, calcium, PTH, and inflammation are positive modulator of FGF23; iron (Fe^{++}) is a negative modulator of FGF23. ASARM, Acidic Serine Aspartate Rich MEPE-Associated Motif; DMP1, dentin matrix protein 1; PHEX, phosphate regulating endopeptidase homolog X-linked; 1,25D, 1,25-dihydroxyvitamin D_3; P, phosphate; Ca, calcium; PTH, parathyroid hormone; Fe^{++}, iron.

However, in recent years, a number of other possible links are emerging between FGF23 and its burden of morbidity and mortality through several pathways [17]. In particular, experimental and clinical evidences point to a specific link between FGF23 and inflammation (Fig. 2). Mechanistic studies have demonstrated that FGF23-mediated activation of FGFR4 is capable of enhancing liver synthesis of inflammatory cytokines [8]. These experimental data are in agreement with the observation that, in CKD patients, serum levels of FGF23 and inflammatory markers (IL-6, TNF, CRP, fibrinogen) are positively correlated [18]. Furthermore, a very recent clinical observation in conservative CKD confirmed the association of serum levels of inflammatory biomarkers and FGF23 with the risk of death, but evidenced that their role is independent and additive [19].

Another biologic effect experimentally acknowledged to FGF23 is the capacity of impairing immune response [9]. In agreement, a re-evaluation of the

HEMO study (originally designed to compare clinical outcomes with different hemodialysis techniques), which included 1,340 hemodialysis patients followed up with repeated biochemical yearly measurements for a median of 3 years, indicate that patients in the highest quartile of FGF23 experienced a higher risk of infectious events, cardiac events, and all-cause mortality [20]. FGF23 has also been shown to associate with atrial fibrillation in CKD. Mehta et al. [21] performed a prospective cohort study on 3,876 individuals demonstrating an independent association of FGF23 with the incidence and prevalence of atrial fibrillation. Mechanistic studies investigating a direct involvement of FGF23 in cardiac arrhythmias however, to date have not been performed. Yet it is likely that the pro-hypertrophic effects of FGF23 promote atrial fibrillation. Fliser et al. [22] assessed FGF23 levels in 227 non-diabetic patients with CKD and followed 177 of the patients prospectively for a median of 53 months to assess the progression of renal disease. In this cohort study, FGF23 independently predicted the progression of CKD after adjustment for covariates such as age, gender, GFR, proteinuria, phosphate, and PTH.

Despite FGF23's powerful correlation with CVD, to date no association has been found between FGF23 and atherosclerosis and arterial calcification. In a study with 1,501 patients, FGF23 did not correlate with coronary artery and thoracic aorta calcium content quantified by computed tomography (CT) [23]. Experimental studies using vascular smooth muscle cells support this hypothesis, since FGF23 also did not induce vascular calcification in vitro [23]. Anemia, a frequent complication seen in patient with CKD and ESRD has also been shown to associate with FGF23. In a cross-sectional observational study of stable CKD patients at stages 3 and 4, high levels of serum FGF23 correlated with low levels of hemoglobin [24]. Certainly, translational studies are needed to elucidate a potential cross-talk between FGF23 and the regulation of iron metabolism and/or the synthesis of erythropoietin.

Due to the growing evidence that FGF23 associates with mortality and morbidity, several studies aimed to reduce FGF23 levels to improve the outcome of CKD patients. Indeed, in a secondary analysis of the EVOLVE trial, the calcimimetic cinacalcet significantly lowered serum FGF23 levels in hemodialysis patients, thereby significantly reducing the rates of cardiovascular events and death. Besides, a novel intravenously administered calcimimetic significantly lowered FGF23 in hemodialysis patients [25]. This study, however, was designed as a non-inferiority trial to target secondary hyperparathyroidism and did not assess clinic outcomes. On the contrary, in a randomized, double blinded, placebo controlled trial, the phosphate binder sevelamer significantly reduced FGF23 in CKD without improving the cardiovascular-related outcomes of interest [26]. Hence, future clinical studies are definitely needed with an intention to reduce FGF23 to clarify if lowering FGF23 improves CVD and overall survival in patients with CKD or ESRD.

FGF23 in Kidney Transplantation

Successful kidney transplantation has been thought to solve the problem of CKD-mineral and bone disorder (CKD-MBD) to a large extent; however, abnormalities in serum calcium, phosphorus, PTH, FGF23, and vitamin D levels are still common in kidney transplant recipients. These conditions are presumably caused by previous bone damage, reduced graft function, CKD-MBD persisting after transplantation or de novo CKD-MBD associated with immunosuppressive therapy.

High levels of PTH, hypophosphatemia, vitamin D deficiency, and hypercalcemia are frequently found in kidney transplant recipients [27, 28]. Interestingly, FGF23 levels decline 3 months after transplantation but remain higher than in CKD patients matched for estimated glomerular filtration rate. Further reductions in FGF23 levels were then repeatedly observed over a longer follow-up period, approximating normal levels 1–3 years after transplantation [28, 29]. In contrast, in a cross-sectional observational study of 279 maintenance kidney recipients with CKD (stages 1–4), Sánchez Fructuoso et al. [30] found that FGF23 levels increased in long-term kidney graft recipients, even in the early stages of CKD, maybe as a result of previous chronic phosphate retention stimulating the secretion of FGF23. These findings support the notion of a persistent (or tertiary) hyperphosphatoninism that in the early post-transplant period mainly reflects previous MBD while in the long-term mirrors a reduction in graft function that is the major determinant of FGF23 serum levels, similar to what is observed in CKD patients.

Nevertheless, only few studies exist investigating the parallel changes of PTH, FGF23 calcium, and phosphate levels in deceased-donor kidney transplantation with a follow-up of >1 year. It is a matter of fact, however, that hyperphosphatoninism, as well as other components of MBD disclose a relapse with the progressive impairment of graft function that follows the progression of chronic allograft deterioration [31, 32].

The impairment of the negative feedback loop between PTH and FGF23, preexistent in ESRD patients, persists after kidney transplantation. While PTH stimulates FGF23 production, at the same time FGF23 suppresses PTH synthesis acting via the klotho–FGFR1 complex in the parathyroid gland or in the absence of klotho through the NFAT pathway [33].

A decreased expression of the klotho-FGFR1 complex in the parathyroid glands causes this parathyroid resistance to FGF23. Speculatively, the clinical use of calcineurin inhibitors that block calcineurin signaling (NFAT pathway) may further increase the susceptibility to develop, or worsen preexisting hyperparathyroidism (HPTH) in patients with reduced klotho expression such as in kidney transplant recipients (KTRs) [33]. At any rate, decreased FGF23 levels within 1 year after transplantation associated with FGF23 resistance and hyperparathyroidism may account for the restored 1,25-dihydroxyvitamin D3 levels within 3–6 months after transplantation [27].

In kidney transplant recipients, FGF23 has emerged as an important mediator of early hypophosphatemia, and its phosphaturic effect is enforced by persistent hyperparathyroidism and 1,25-dihydroxyvitamin D_3 deficiency. In the long-term, hypophosphatemia and renal phosphate loss are mainly related to persistent hyperparathyroidism. In the first year after transplantation, the faster normalization of FGF23 levels, compared with PTH levels, suggests that FGF23-producing cells may experience a quick reset of their activities in response to the recovery of renal function, all despite persistently elevated PTH levels which stimulate FGF23 production [34]. The long-lasting PTH-related renal phosphate wasting in kidney transplant recipients induces a negative phosphate balance that is appropriately sensed by osteocytes, thus decreasing the FGF23 production in an effort to conserve phosphate, despite the stimulatory effects of high PTH levels (Fig. 3).

Few studies have analyzed the changes in FGF23, PTH, and phosphate levels in living-donor kidney transplants. These studies agree in describing a faster normalization of FGF23 and hypophosphatemia within 1 year after transplantation and a reduced prevalence of hyperparathyroidism [28]. Two factors are held accountable for these findings: the shorter vintage dialysis prior to transplantation, associated with a less severe degree of CKD-MBD and the faster normalization of renal function in living-donor kidney recipients.

Chronically elevated FGF23 levels may be considered ultimately as a maladaptive process in patients with CKD, given the strong associations between higher FGF23 levels and increased risk of LVH, congestive heart failure, CKD progression, and death. To date, clinical studies analyzing the role of FGF23 in transplant patients, however, are still missing.

In a prospective study of stable kidney transplant recipients, elevated FGF23 levels were independently associated with an increased risk of cardiovascular and all-cause mortality and allograft loss [14, 35].

There are several possible mechanisms that may explain this finding. In vitro and in vivo studies have shown that 1,25-dihydroxyvitamin D_3 decreases T cell activation and proliferation and inhibits dendritic cell differentiation and matu-

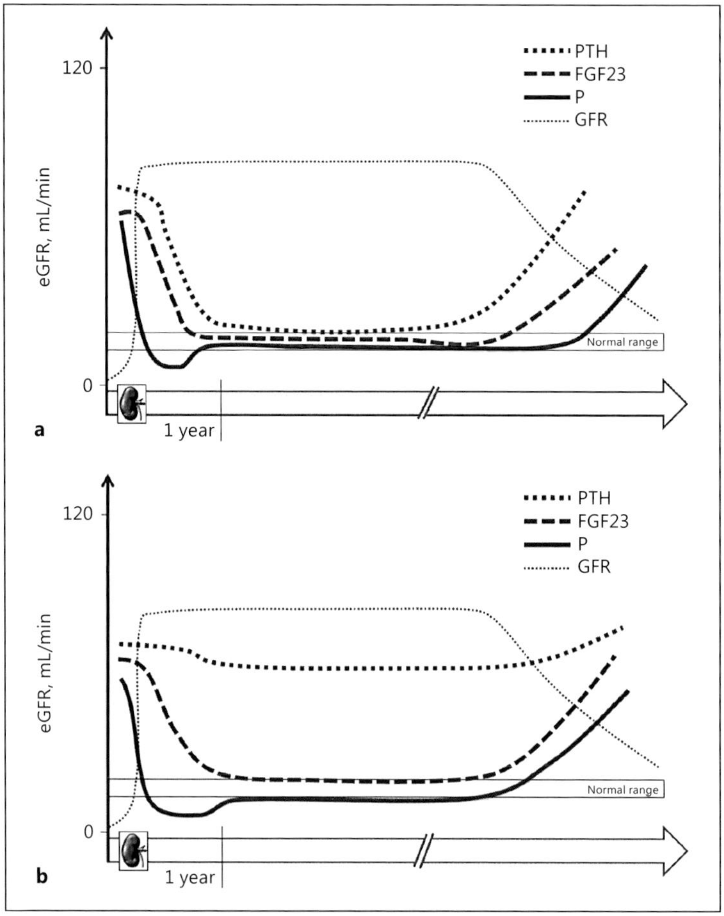

Fig. 3. Trends of PTH, FGF23, and phosphate (P) levels at 1 year after renal transplantation and in late period (follow-up). Hyperphosphatoninism in the early post-transplant period mainly reflects previous mineral and bone disorders while in the long-term mirrors graft function and is the main mediator of early hypophosphatemia. In the long term, along with deterioration of graft function, hyperphosphatoninism, and hyperparathyroidism relapse to maintain normophosphatemia and normocalcemia (**a**). "Persistent" hyperparathyroidism results from pre-existing CKD-MBD with secondary hyperparathyroidism likely complicating post-transplant follow-up with hypophosphatemia and hypercalcemia without affecting the normalization of FGF23. At any rate, hyperphosphatoninism relapses along with the progressive impairment of graft function (**b**).

ration, while its supplementation may have beneficial effects on chronic allograft nephropathy [27, 36]. FGF23-mediated suppression of 1,25-dihydroxyvitamin D_3 is one possible mechanism through which high FGF23 levels could contribute to allograft loss. In kidney transplant recipients, phosphate depletion, in conjunction with high PTH levels, vitamin D deficiency and chronic steroid use, might worsen skeletal demineralization and contribute directly to fractures, which in turn could increase the risk of mortality. High levels of FGF23 may impair neutrophil recruitment, which additionally could jeopardize antibacterial defense in an already immunocompromised setting [9]. FGF23 has also been shown to cause inflammation and this particular effect may significantly contribute to the inflammatory burden of renal transplant patients [37]. Moreover, chronic elevated FGF23 levels could mimic the known effects of FGF2 inducing glomerulosclerosis and thereby directly contribute to chronic allograft nephropathy and death [7, 28].

Of note, in a retroactive study on deceased pediatric CKD patients, kidney transplantation has been shown to lower cardiac expression of FGF23 and FGFR4 and to normalize klotho protein levels [15].

Finally, an underestimated issue is the effect of immunosuppressive drugs on FGF23 levels, considering their links with vitamin D metabolism and PTH synthesis [27].

Taken together, clearly clinical and mechanistic studies are needed to further elucidate the role of FGF23 in the setting of renal transplantation.

Future Perspectives

FGF23 seems to function as a circulating factor that can directly contribute to cardiac hypertrophy, inflammation, and impaired host response in CKD. Further detailed mechanistic studies are needed to identify a potential causative involvement of FGF23 in further common CKD-related pathophysiologic alterations such as anemia, progression of renal disease, endothelial dysfunction, and atherosclerosis as well other end organ-specific diseases associated with systemic inflammation. Besides, future clinical studies could emphasize the potential differences in gender, ethnicity, and genetic characteristics. Finally, novel laboratory methods have to be developed to reliably assess klotho levels in patients as well as klotho and FGF23 activity in vitro and in experimental in vivo settings. Extending our knowledge of FGF23-FGFR biology in the context of CKD and systemic inflammation will eventually lead to the identification of novel drug targets and the development of pharmacological interventions that ultimately might reduce the burden of cardiovascular injury, decrease systemic inflammation, prevent infections, prolong kidney transplant survival, and ultimately decrease mortality among the millions of CKD patients worldwide.

References

1 Ben-Dov IZ, Galitzer H, Lavi-Moshayoff V, Goetz R, Kuro-o M, Mohammadi M, et al: The parathyroid is a target organ for FGF23 in rats. J Clin Invest 2007;117:4003–4008.

2 Shimada T, Kakitani M, Yamazaki Y, Hasegawa H, Takeuchi Y, Fujita T, et al: Targeted ablation of Fgf23 demonstrates an essential physiological role of FGF23 in phosphate and vitamin D metabolism. J Clin Invest 2004; 113:561–568.

3 Shimada T, Hasegawa H, Yamazaki Y, Muto T, Hino R, Takeuchi Y, et al: FGF-23 is a potent regulator of vitamin D metabolism and phosphate homeostasis. J Bone Miner Res 2004;19:429–435.

4 Andrukhova O, Smorodchenko A, Egerbacher M, Streicher C, Zeitz U, Goetz R, et al: FGF23 promotes renal calcium reabsorption through the TRPV5 channel. EMBO J 2014; 33:229–246.

5 David V, Martin A, Isakova T, Spaulding C, Qi L, Ramirez V, et al: Inflammation and functional iron deficiency regulate fibroblast growth factor 23 production. Kidney Int 2016;89:135–146.

6 Faul C, Amaral AP, Oskouei B, Hu MC, Sloan A, Isakova T, et al: FGF23 induces left ventricular hypertrophy. J Clin Invest 2011; 121:4393–4408.

7 Grabner A, Amaral AP, Schramm K, Singh S, Sloan A, Yanucil C, et al: Activation of cardiac fibroblast growth factor receptor 4 causes left ventricular hypertrophy. Cell Metab 2015;22:1020–1032.

8 Singh S, Grabner A, Yanucil C, Schramm K, Czaya B, Krick S, et al: Fibroblast growth factor 23 directly targets hepatocytes to promote inflammation in chronic kidney disease. Kidney Int 2016;90:985–996.

9 Rossaint J, Oehmichen J, Van Aken H, Reuter S, Pavenstädt HJ, Meersch M, et al: FGF23 signaling impairs neutrophil recruitment and host defense during CKD. J Clin Invest 2016;126:962–974.

10 Vervloet MG, van Zuilen AD, Heijboer AC, ter Wee PM, Bots ML, Blankestijn PJ, et al: Fibroblast growth factor 23 is associated with proteinuria and smoking in chronic kidney disease: an analysis of the MASTERPLAN cohort. BMC Nephrol 2012;13:20.

11 Ravikumar P, Ye J, Zhang J, Pinch SN, Hu MC, Kuro-o M, et al: α-Klotho protects against oxidative damage in pulmonary epithelia. Am J Physiol Lung Cell Mol Physiol 2014;307:L566–L575.

12 Gutiérrez OM, Mannstadt M, Isakova T, Rauh-Hain JA, Tamez H, Shah A, et al: Fibroblast growth factor 23 and mortality among patients undergoing hemodialysis. N Engl J Med 2008;359:584–592.

13 Isakova T, Xie H, Yang W, Xie D, Anderson AH, Scialla J, et al: Fibroblast growth factor 23 and risks of mortality and end-stage renal disease in patients with chronic kidney disease. JAMA 2011;305:2432–2439.

14 Wolf M, Molnar MZ, Amaral AP, Czira ME, Rudas A, Ujszaszi A, et al: Elevated fibroblast growth factor 23 is a risk factor for kidney transplant loss and mortality. J Am Soc Nephrol 2011;22:956–966.

15 Leifheit-Nestler M, Große Siemer R, Flasbart K, Richter B, Kirchhoff F, Ziegler WH, et al: Induction of cardiac FGF23/FGFR4 expression is associated with left ventricular hypertrophy in patients with chronic kidney disease. Nephrol Dial Transplant 2016;31: 1088–1099.

16 Scialla JJ, Wolf M: Roles of phosphate and fibroblast growth factor 23 in cardiovascular disease. Nat Rev Nephrol 2014;10:268–278.

17 Mazzaferro S, Pasquali M, Pirrò G, Rotondi S, Tartaglione L: The bone and the kidney. Arch Biochem Biophys 2010;503:95–102.

18 Munoz Mendoza J, Isakova T, Ricardo AC, Xie H, Navaneethan SD, Anderson AH, et al: Fibroblast growth factor 23 and Inflammation in CKD. Clin J Am Soc Nephrol 2012;7: 1155–1162.

19 Munoz Mendoza J, Isakova T, Cai X, Bayes LY, Faul C, Scialla JJ, et al: Inflammation and elevated levels of fibroblast growth factor 23 are independent risk factors for death in chronic kidney disease. Kidney Int 2017;91: 711–719.

20 Chonchol M, Greene T, Zhang Y, Hoofnagle AN, Cheung AK: Low vitamin D and high fibroblast growth factor 23 serum levels associate with infectious and cardiac deaths in the HEMO study. J Am Soc Nephrol 2016;27: 227–237.

21 Mehta R, Cai X, Lee J, Scialla JJ, Bansal N, Sondheimer JH, et al: Association of fibroblast growth factor 23 with atrial fibrillation in chronic kidney disease, from the chronic renal insufficiency cohort study. JAMA Cardiol 2016;1:548–556.
22 Fliser D, Kollerits B, Neyer U, Ankerst DP, Lhotta K, Lingenhel A, et al: Fibroblast growth factor 23 (FGF23) predicts progression of chronic kidney disease: the Mild to moderate kidney disease (MMKD) study. J Am Soc Nephrol 2007;18:2600–2608.
23 Scialla JJ, Lau WL, Reilly MP, Isakova T, Yang HY, Crouthamel MH, et al: Fibroblast growth factor 23 is not associated with and does not induce arterial calcification. Kidney Int 2013;83:1159–1168.
24 Tsai MH, Leu JG, Fang YW, Liou HH: High fibroblast growth factor 23 levels associated with low hemoglobin levels in patients with chronic kidney disease stages 3 and 4. Medicine (Baltimore) 2016;95:e3049.
25 Block GA, Bushinsky DA, Cheng S, Cunningham J, Dehmel B, Drueke TB, et al: Effect of etelcalcetide vs cinacalcet on serum parathyroid hormone in patients receiving hemodialysis with secondary hyperparathyroidism: a randomized clinical trial. JAMA 2017;317: 156–164.
26 Chue CD, Townend JN, Moody WE, Zehnder D, Wall NA, Harper L, et al: Cardiovascular effects of sevelamer in stage 3 CKD. J Am Soc Nephrol 2013;24:842–852.
27 Cianciolo G, Galassi A, Capelli I, Angelini ML, La Manna G, Cozzolino M: Vitamin D in kidney transplant recipients: mechanisms and therapy. Am J Nephrol 2016;43:397–407.
28 Cianciolo G, Cozzolino M: FGF23 in kidney transplant: the strange case of Doctor Jekyll and Mister Hyde. Clin Kidney J 2016;9:665–668.
29 Evenepoel P, Meijers BK, de Jonge H, Naesens M, Bammens B, Claes K, et al: Recovery of hyperphosphatoninism and renal phosphorus wasting one year after successful renal transplantation. Clin J Am Soc Nephrol 2008; 3:1829–1836.
30 Sánchez Fructuoso AI, Maestro ML, Pérez-Flores I, Valero R, Rafael S, Veganzones S, et al: Serum level of fibroblast growth factor 23 in maintenance renal transplant patients. Nephrol Dial Transplant 2012;27:4227–4235.
31 Scolari MP, Cappuccilli ML, Lanci N, La Manna G, Comai G, Persici E, et al: Predictive factors in chronic allograft nephropathy. Transplant Proc 2005;37:2482–2484.
32 Riella LV, Djamali A, Pascual J: Chronic allograft injury: mechanisms and potential treatment targets. Transplant Rev (Orlando) 2016;31:1–9.
33 Olauson H, Lindberg K, Amin R, Sato T, Jia T, Goetz R, et al: Parathyroid-specific deletion of Klotho unravels a novel calcineurin-dependent FGF23 signaling pathway that regulates PTH secretion. PLoS Genet 2013; 9:e1003975.
34 Wolf M, Weir MR, Kopyt N, Mannon RB, Von Visger J, Deng H, et al: A prospective cohort study of mineral metabolism after kidney transplantation. Transplantation 2016; 100:184–193.
35 Baia LC, Humalda JK, Vervloet MG, Navis G, Bakker SJ, de Borst MH: Fibroblast growth factor 23 and cardiovascular mortality after kidney transplantation. Clin J Am Soc Nephrol 2013;8:1968–1978.
36 Cianciolo G, La Manna G, Cappuccilli ML, Lanci N, Della Bella E, Cuna V, et al: VDR expression on circulating endothelial progenitor cells in dialysis patients is modulated by 25(OH)D serum levels and calcitriol therapy. Blood Purif 2011;32:161–173.
37 La Manna G, Cappuccilli ML, Cianciolo G, Conte D, Comai G, Carretta E, et al: Cardiovascular disease in kidney transplant recipients: the prognostic value of inflammatory cytokine genotypes. Transplantation 2010;89: 1001–1008.

Alexander Grabner, MD
Duke University Medical Center
Department of Medicine, Division of Nephrology
2 Genome Court, MSRB II, Durham NC 27705 (USA)
E-Mail alexander.grabner@duke.edu

Perspectives in Nephrology and Chronic Kidney Disease

La Manna G, Ronco C (eds): Current Perspectives in Kidney Diseases.
Contrib Nephrol. Basel, Karger, 2017, vol 190, pp 96–107 (DOI: 10.1159/000468954)

Restenosis in Hemodialytic Fistulas and Chronic Kidney Disease-Associated Vascular Disease: Two Pathologies Driven by Metakaryotic Stem Cells

Gianandrea Pasquinelli[a] · William G. Thilly[d] · Elena V. Gostjeva[d] · Paola Todeschini[a] · Giuseppe Cianciolo[a] · Claudio Ronco[b, c] · Gaetano La Manna[a]

[a]Department of Experimental Diagnostic and Specialty Medicine (DIMES), S. Orsola-Malpighi Hospital, University of Bologna, Bologna, [b]Department of Nephrology, Dialysis and Transplantation, and [c]International Renal Research Institute of Vicenza, San Bortolo Hospital, Vicenza, Italy; [d]Laboratory in Metakaryotic Biology (LIMB), Department of Biological Engineering, Massachusetts Institute of Technology, Cambridge, MA, USA

Abstract

Chronic kidney disease (CKD) exacerbating vascular disease poses a major challenge to nephrology. Surgically placed vascular fistulas, as an aid to hemodialysis prior to kidney transplant, have extended many lives, while post-surgical restenosis closure of the fistula by smooth muscle cells affects many lives. When post-surgical restenosis is developed, palliative measures are almost always surgical: there are no effective drug treatments. In this study, we offer a testable hypothesis that effects of CKD on widely distributed vascular diseases and the phenomenon of fistula restenosis are both driven by the pathologic creation of non-dividing smooth muscle cells via asymmetric division of exponentially increasing metakaryotic stem cells. In slow growing atherosclerotic plaques, the Benditts demonstrated clonality of smooth muscle cells that we posit originate in a single mutated metakaryotic stem cell of fetal/juvenile vasculogenesis. In the fast process of fistula restenosis, we posit quiescent metakaryotic stem cells "on call" for wound healing among which are rare stem cells that have lost the ability to cease division. These hypotheses and suggestions for specific research paths toward development of effective drug therapies are built on (a) our shared discoveries of the role of metakaryotic stem cells in organogenesis, carcinogenesis, and atherosclerotic plaque formation and (b) the recent finding that metakaryotic cancer stem cells are constitutively resistant to radio- and chemotherapies yet sensitive to killing by a wide range of existing drugs. We propose to test these hypotheses in discarded fistulas and stem cells derived therefrom, and, if supported, to test drug-eluting devices to block fistula restenosis.

Introduction

Patients with chronic kidney disease (CKD) have a markedly increased incidence of cardiovascular events and cardiovascular deaths compared to age-matched counterparts in the general population, regardless of the stage of disease [1, 2].

Vascular disease in this population is undoubtedly multifactorial, as there is a high prevalence of traditional cardiovascular risk factors, but it also has its own specific characteristics. Among the key determinants of CKD vascular disease, arterial impairment of vasodilation in response to various stimuli including increased shear stress has been considered an endothelial feature of dysfunction [3]. Impairment of the balance between endothelium-derived relaxing, that is, nitric oxide, and contracting factors, that is, endothelin 1 and angiotensin-II, predisposes the vasculature to vasoconstriction, leukocyte adherence, platelet activation, increased cellularity, pro-oxidation, thrombosis, impaired coagulation, vascular inflammation, and atherosclerosis [4].

In addition, there is a growing body of evidence that disturbances of mineral metabolism may act as a trigger of cardiovascular disease (CVD) in CKD patients. The synthetic definition of Chronic Kidney Disease Mineral and Bone Disorder (CKD-MBD) includes the concept that the indexes of mineral metabolism act on the vascular wall either by pleiotropic mechanisms [5, 6] or by enhancing pathways involved in vascular injury [7] or in the perturbation of bone vascular axis [8].

Further, an increased arterial stiffness is reported. Stiffness is determined, in large part, by the elastin to collagen ratio in their walls [9]; ageing leads to progressive arterial stiffness as a result of elastic fiber degeneration and atherosclerotic development. In CKD, the increased arterial stiffness is also a consequence of media calcification, a feature that is prominent in this patient population, and to intima calcification related to end-stage atherosclerosis [10]. Nevertheless, the mechanisms underlying the increased incidence of CVD in CKD patients remain incompletely understood.

Despite pharmacological and technological advances in renal replacement therapy [11], people with CKD continue to die from CVD. A key role is played by the vascular access which represents an important contributor to the micro inflammatory status of dialysis patients, triggering an inflammatory and immunological reaction as well as oxidative stress activation. These conditions result in an increased synthesis of pro-inflammatory mediators (i.e., C-reactive protein [CRP], IL-6, TNF-α), which have been associated with all-cause and cardiovascular mortality [12]. In particular, patients undergoing regular hemodialysis using surgically placed arteriovenous fistulas (AVF) require frequent replacement of fistulas due to restenotic blockage [13]. Intimal hyperplasia (IH) is considered the culprit lesion in failed fistulas, resulting in luminal narrowing and stenosis. This finding is an intriguing issue, since IH is also the earlier lesion in atherosclerotic process and smooth muscle cells are involved in both processes.

Fistula maturation might be partially hindered by pre-existent vascular changes, often present in patients with CKD. Indeed, recent studies showed marked pre-existing IH in venous segments of patients with end-stage renal disease prior to vascular access surgery [14].

Vascular Stem Cells Dysfunction and CKD-Associated Vascular Disease

The human arterial wall is believed to contain multiple classes of resident stem cells that together maintain vascular homeostasis during a lifetime [15]. Such stem cell populations have been isolated and characterized in different human vascular adult tissues such as aorta and femoral artery [16, 17], pulmonary artery [18], internal mammary artery [19], and saphenous veins of the leg [20, 21].

Unlike stem cells colonic crypts that divide by asymmetric mitoses about every 128 days, vascular stem cells are commonly quiescent having a proliferative activity that is strongly induced by injury. As such they belong to the class sometimes designated as "wound healing" stem cells that are possibly comprised of amitotic non-eukaryotic stem cells discussed below. After the vascular wall wound has been repaired or regenerated, these cells are believed to enter a non-proliferative state, but it is possible they are used up in terminal extinction divisions. Mechanisms regulating these processes are complex and possibly interact with inflammatory conditions attendant upon wounding [22].

Renal impairment is particularly associated with chronic low-grade inflammation with elevated levels of CRP, IL-2, IL-6, INF-gamma, TNF-alpha, sIL-2R, ICAM-1, and VCAM-1 [23, 24] that could well perturb the stem cell functional activity in the vascular wall. It is thus possible that the high incidence of vascular disease in CKD is determined by chronic inflammatory stress on vascular stem cells that disrupts vascular homeostasis. Documented examples of such possibilities are hereinafter reported.

Mesenchymal stem cells that reside in calcified and inflamed abdominal aorta aneurysms of patients with high serum levels of MMP-9 protein display high osteogenic potential and are prone to perturbed angiogenic ability. Both features indicate progression of advanced vascular disease. Their osteogenic commitment is strongly potentiated when co-cultured with PHA stimulated peripheral blood mononuclear cells or inflammatory cytokines [25]. Similarly, vascular smooth muscle cells have been interpreted to a shift to an osteogenic phenotype when subjected to oxidative stress, oxylipids, and phosphates [26]. These interpretations may require revision as the role of vascular stem cells in osteoblast formation becomes clearer.

Uncharacterized cells associated with smooth muscle cells produce a collagenous matrix in which calcium- and phosphorus-rich matrix vesicles are generated as nucleation sites for hydroxyapatite in the extracellular matrix [27].

A further example of the vascular stem cell disturbance is provided by Orrico et al. [28]. They reported in diabetic patients with CKD, peripheral artery obstructive disease and critical limb ischemia, decreased neoangiogenesis in the media ostensibly as a result of functional impairment of circulating and resident endothelial progenitors.

Vascular Stem Cells Over-Proliferation and IH

A yet unaddressed issue certainly concerns the ways that vascular wall resident stem cells respond to tissue damage and proliferate according to tissue healing requirements. But just as important are the processes by which they cease and avoid over-proliferation. For example, IH is the pathologic lesion in AVFs that may result in stenosis and ultimately thrombosis [14]. It is characterized primarily by concentric intima expansion of vein graft by an α-smooth muscle actin (α-SMA) positive cell population (Fig. 1a–c). The vast majority of α-SMA positive cells in the intimal lesions exhibit a synthetic smooth muscle cell phenotype, which is quite similar to that of myofibroblasts described in wound healing. These cells, derived from proliferation and dedifferentiation of contractile smooth muscle cells (Figure 1d), unexpectedly fail to stain with the universal proliferation marker ki-67. As anticipated in the previous section, the modified smooth muscle cells might arise directly from stem cells resident in the vascular wall. Accordingly, synthetic smooth muscle cells may comprise of wound healing stem cells, created continuously during organogenesis. If so, these could be related in nature to the non-eukaryotic, amitotic stem cells, named metakaryotic cells, each typically observed displaying a large, hollow bell-shaped nucleus previously considered to be forms of terminally differentiated "goblet cells" [29]. Metakaryotes, discussed in the next section, were discovered in all fetal tissues, first in a tubular syncytial form (4–12 weeks gestation) and afterwards till maturity in mononuclear form, and in juveniles and

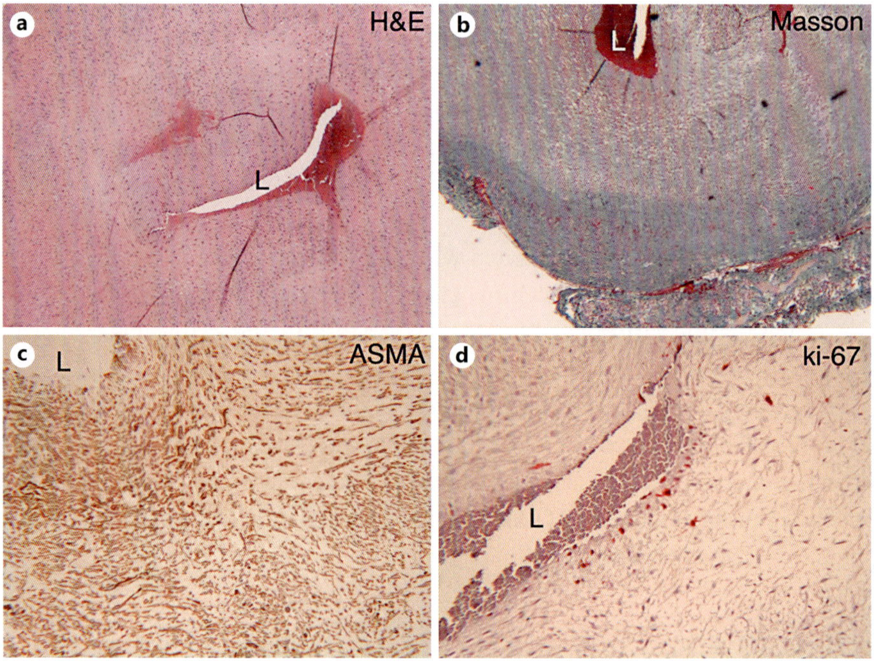

Fig. 1. Conventional histopathology of arteriovenous fistulas removed due to intimal hyperplasia. **a** At low magnification, the vascular lumen (L) is dramatically constricted by a concentric proliferation of spindle-shaped cells (H&E), that (**b**) are embedded in loose mucoid matrix (Masson trichrome stain); (**c**) the proliferated cells express α-smooth muscle actin (ASMA) intensely, but (**d**) fail to stain with ki-67 universal proliferation marker. **a**, **b** Original magnification: 45×; **c**, **d** original magnification: 100×.

adults scattered within the mesenchyme generally and as we have found throughout the adult arteries [30].

Another possibility is that multipotent circulating stem cells supervise their generation. In fact under normal and pathological conditions, there is a continuous egress out of the bone marrow into the bloodstream of stem cells, including those committed towards specific lineages that participate in the maintenance of the cardiovascular integrity.

Atherosclerosis and diabetes, co-morbidities often associated with CKD, are accompanied by multilineage depletion of circulating vascular stem cells [31]. Considering that PTH and osteoblasts play the role of pivotal directors of stem cell mobilization from bone marrow, any change in bone remodeling inherent to the CKD-MBD could influence the circulating stem cell egress. This condition, known as "mobilopathy", can persist until the appearance of progenitors or stem cell subsets endorsed by an osteogenic phenotype [32].

Metakaryotic Stem Cells

Metakaryotic cells appear to be the primary form of organogenic stem cell common to plants and animals, indicating an evolutionary origin prior to the separation of these phyla [33]. They were first discovered by one of us (E.V.G.) in early human fetal development of the colon [29, 34]. Her findings were then extended to other human organs in development as well as to pathologic states such as colonic adenomas, adenocarcinomas, and recently, metastases (unpublished results). In particular, they appear to be the stem cell candidate responsible for the generation of synthetic smooth muscle cells seen in atherosclerotic plaque, and seem probable in post-surgical restenosis.

Metakaryotes have morphologic and physiologic characteristics that distinguish them from eukaryotic cells. Eukaryotic nuclei have a wide variety of shapes but they are enclosed in a nuclear membrane that resides within the cytoplasm. Metakaryotic cells have hollow, bell-shaped nuclei that are appended to the cytoplasm rather than contained within it. Eukaryotic human cell genomes in interphase are arranged as unpaired chromatids with physically distinct ends or "telomeres". In metakaryotic cells, interphase genomes appear to be a varying set of circular chromosomes comprised of paired maternal/paternal chromatids joined at the centromeres and both telomeres [35]. Multiple chromatid pairs may co-exist in one circular chromosome joined at their telomeric regions. Eukaryotic cells copy their dsDNA genomes to make 2 dsDNA copies during S-phase, which are segregated into paired chromatids by mitosis. Metakaryotic cells copy the entire dsDNA genome into a dsRNA/DNA replicative intermediate that is segregated during an amitotic process and returned to the dsDNA form after segregation into sister nuclei [36]. This process is used both in symmetric and asymmetric amitotic divisions. The forms of amitotic figures vary. Most commonly, the symmetric segregation resembles the separation of 2 stacked paper cups [29]. In another form, symmetrical amitosis resembles a sphere bifurcated and separated as 2 equal hemispheres [33]. The many eukaryotic nuclear forms in organogenesis and carcinogenesis originate in asymmetric amitoses of a metakaryotic parent cell. After segregation from the metakaryotic parent, such newly minted eukaryotic cells may undergo a finite number of mitotic doublings ranging from zero (smooth muscle cells) to as many as 13 (colonic adenocarcinoma crypts) [37]. If the metakaryotic cytoplasm is an extended oblate spheroid with the nucleus attached on one end, it is described as a "goblet cell." If the metakaryotic cell has a spherical cytoplasm, it is described as a "signet ring cell." As goblet and signet ring cells were not observed to undergo mitoses, it was assumed that they were non-dividing cells and thus considered "terminally differentiated" [33].

The curious mode of dsDNA genome replication in metakaryotic cells seems to have endowed them with very high rates of genetic mutation. Measurements of the distribution of mutant colony sizes in microanatomical samples of human lungs and colons indicated that metakaryotic cells during development have mutation rates about a thousand and a hundred fold higher than that found in human eukaryotic near-diploid lymphoblasts or embryonic stem cells grown in culture [38]. This high rate of mutation suggests a role for metakaryotic cells in the changes in stem cells required for "initiation" in carcinogenesis and probably in the origination of clonal diseases such as atherosclerosis in general. In fact an algebraic model of colorectal carcinogenesis was built upon the observed rate of mutational loss of the APC gene in colonic stem cells prior to maturity, the mathematical expectations were not distinguishable from the observed mortality rate for the diseases recorded in European American males born between 1890 and 1909 throughout their adult life spans of 15–19 through 100–104 years [39].

As our Italian-American collaboration has observed similar behavior of metakaryotic cells associated with increased smooth muscle cell numbers and, even osteoblasts, in atherosclerotic plaques, we naturally looked at the processes of post-surgical restenosis with several collaborators and espied what appears to be creation of smooth muscle cells by asymmetric divisions of metakaryotic cells [33]. These observations prompted us to offer this paper incorporating the facts as we presently understand restenotic cell dynamics so that others may consider their value.

Figure 2 offers some images that may guide those who look for metakaryotes in different settings of organogenesis or pathogenesis. It shows several examples of metakaryotes in vascular settings giving rise to single nuclei of smooth muscle cells by asymmetric amitoses (Fig. 2b–e) [33]. It also offers images reflecting ways to observe the dsRNA/DNA pangenomic replicative intermediate used by these bizarre stem cells [36] and the fact that they, like eukaryotic cells, express the ki-67 gene when involved in genomic replication.

New and Emerging Histopathology: High-Resolution Microscopy and Cell Culture Techniques

In depth histopathological analysis of post-surgical restenotic tissues has yet to comprehend the nature or number of stem cells responsible for the basic pathology. If native fistular restenotic plaques could be removed surgically and analyzed, much could be learned. There are limited observations on the fact that the vascular post-surgical restenotic tissues in a majority of cases are not removed

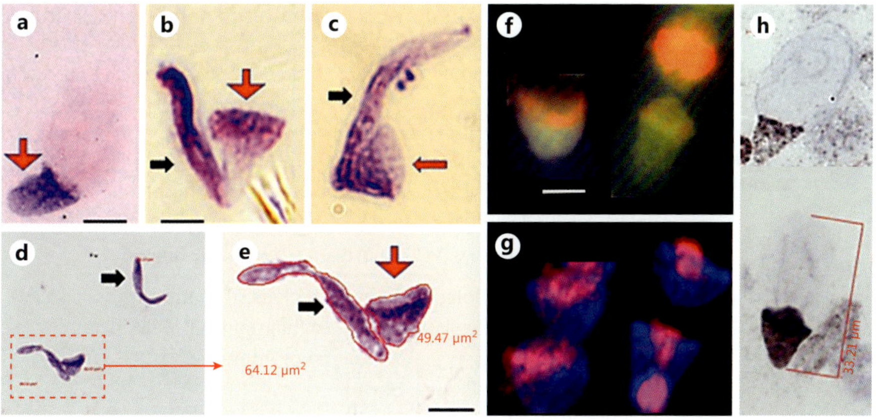

Fig. 2. Metakaryotic stem cells with bell-shaped nuclei from human tissues. **a** Quiescent nucleus from calcified human aortic valve; Feulgen purple dsDNA stain; **b–e** asymmetrically dividing metakaryotes (red arrows) showing "tobacco pipe" form: asymmetric division results in the spindle-shaped nucleus of a smooth muscle cell (black arrows) from a human calcified carotid artery atherosclerotic plaque; **f** acridine orange staining in dividing metakaryotes after RNAse treatment, ssDNA, orange color; dsDNA, green color; **g** dsRNA/DNA antibody IF staining (rhodamine red, counterstained with DAPI blue) in symmetrically (left image) and asymmetrically (right image) dividing human metakaryotes; **h** Ki-67 IHC staining showing negative staining in quiescent (upper image) and positive staining in asymmetrically (bell- to oval shaped-nucleus) dividing metakaryotic nucleus. Scale bar = 5 microns.

in medical practice. An opportunity is offered, however, by the practice of removing the failed AVF employed in hemodialysis, which could be prepared for histologic examination immediately upon surgical removal as discarded tissue. The processes described in part for fistulas seem similar to those described in peripheral and cardiac vessel restenosis. It seems to us that failed fistulas offer success in the study of vascular post-surgical restenosis. Progress in understanding the metakaryotic stem cell physiology in pathological vascular remodeling and vascular remodeling at large will, we think, require some technical advancement. In Table 1, we cite some new, or at least not widely used, including high resolution light/fluorescent microscopy, methods of preparing samples and means to dissociate tissues without enzymatic digestion.

The use of extended periods of tissue immersion in cold Ca^{++} and Mg^{++} free cold phosphate buffer seems to break down what would appear to be cold-labile adherence structures among cells of many tissues and pathologic lesions. As the cells derived are generally able to grow post cold incubation, a means to grow metakaryotic stem cells directly from discarded failed fistulas is worth trying. In this effort, we recommend the use of CO_2 and antibiotic free conditions, prefer-

Table 1. Imaging of metakaryotic stem cells in CKD-related vascular disease and hemodialysis fistulas

Histopathological method	Advantages of application in CKD vascular disease	Disadvantages of application in CKD vascular disease	References
Post-fixation enzymatic tissue dissociation	Single, intact cells and stem cells larger than 5 microns could be visualized in remodeled, calcified vascular lesions	Cell surface antigens, like most of CD stem cell markers, cannot be visualized following enzymatic digestion	[29, 30, 33]
Unfixed tissue; non-enzymatic tissue dissociation: cold, Ca^{++}, Mg^{++} free, phosphate buffer denaturation of cold-labile molecular associations	Cell membrane-bound molecules are preserved, CD markers are visualized in intact cells and stem cells; live dissociated tissue cells available for primary cell culture of vascular stem cells	Involves strict time limits to maintain cell viability; labor intensive	[40, 41]
Unfixed tissue: tissue expansion in polyelectrolyte gels (expansion microscopy)	Vascular lesion could be visualized in ~100 micron sections. 3D imaging of entire tissue compartment and intact cells; cellular structures, <1 micron, like microvesicles, could also be visualized	Limitations in isotropic expansion of rigid and calcified vascular tissue; supra-cellular structures such as tissue turnover units larger than 60 microns might not be detected	[42]

ably with media constituents such as amino acids as are found in human plasma [37]. These are the conditions used to study the general toxicology of human metakaryotic cancer stem cells.

Using these conditions we were able to discover, as was oft suspected for the then unknown forms of tumor stem cells, that metakaryotic stem cells in tumors and cell cultures are constitutively resistant to killing by X-rays and chemotherapeutic drugs in common use [37]. In contrast, we found metakaryotic cells in human adenocarcinoma-derived cultures (HT-29 cells) to be [37]. Next we discuss the clinical studies necessary to discover if treatments well tolerated for such drugs actually kill metakaryotic stem cells in vivo. Would it not seem prudent to culture the metakaryotic stem cells directly from vascular fistulas that showed restenosis to discover what drugs might be toxic to them under conditions expected to be well tolerated by patients? Do not the peculiar modes of using fistulas permit one to imagine an insert that, after a period for primary wound healing, would slowly elute metakaryocidal drugs locally to extend the useful lifetime of these already life-extending surgical entities? There is much to be done but at least the hypotheses that may now be offered are specific: metakaryotic cells drive the increase in the number of smooth muscle cells and osteoblastic cells in vascular fistulas and can be killed or retarded by the use of common drugs with metakaryocidal activities that could be co-implanted with fistulas and maintain a condition preventing the pathologic growth of stem cells for periods appreciably longer than now possible.

Conclusions

Despite many efforts, we do not know yet what mechanisms give rise to the mass of α-SMA positive spindle cell population representing the culprit of CKD-associated vascular disease. Such cells, commonly described as synthetic smooth muscle cells, are crucial players in the development of atherosclerotic plaque, in stent-restenosis and IH of failed AVF. They represent a cell population sharing strict homologies with the myofibroblasts seen temporarily in wound healing. It is noteworthy that these cells do not exist in normal conditions but may be observed during organogenesis. Further their lesional increase in number was always considered a consequence of cell proliferation; however, this does not seem to be the case. Synthetic smooth muscle cells may be derived from amitotic expansion of wound healing stem cell, an organogenic stem cell common to plants and animals also known as metakaryotic stem cells.

This hypothesis can be tested by applying emerging histopathology techniques to discarded AVF removed due to IH. Using ex vivo culture conditions, we expect to discover that wound-healing stem cells could be more sensitive than eukaryotes in killing a wide range of existing drugs, thus paving the way for therapeutic strategies expected to prevent the burden of vascular disease in CKD population.

Acknowledgments

W.G.T. and E.V.G. gratefully acknowledge the generous support of the United Therapeutics Corporation, Silver Spring, MD, USA during the period 2007–2014. Research in their MIT laboratory is now supported by their private funds.

References

1 Navaneethan SD, Schold JD, Arrigain S, et al: Cause-specific deaths in non-dialysis-dependent CKD. J Am Soc Nephrol 2015;26:2512–2520.
2 Wen CP, Matsushita K, Coresh J, et al: Relative risks of chronic kidney disease for mortality and end-stage renal disease across races are similar. Kidney Int 2014;86:819–827.
3 Davignon J, Ganz P: Role of endothelial dysfunction in atherosclerosis. Circulation 2004;109:27–32.
4 Verma S, Anderson TJ: Fundamentals of endothelial function for the clinical cardiologist. Circulation 2002;105:546–549.

5 Cozzolino M, Ureña-Torres P, Vervloet MG, et al: Is chronic kidney disease-mineral bone disorder (CKD-MBD) really a syndrome? Nephrol Dial Transplant 2014;29:1815–1820.
6 Cianciolo G, La Manna G, Cappuccilli ML, et al: VDR expression on circulating endothelial progenitor cells in dialysis patients is modulated by 25(OH)D serum levels and calcitriol therapy. Blood Purif 2011;32:161–173.
7 Hanudel M, Jüppner H, Salusky IB: Fibroblast growth factor 23: fueling the fire. Kidney Int 2016;90:928–930.
8 Cianciolo G, La Manna G, Donati G, et al: Effects of unfractioned heparin and low-molecular-weight heparin on osteoprotegerin and RANKL plasma levels in haemodialysis patients. Nephrol Dial Transplant 2011;26: 646–652.
9 Anderson TJ: Arterial stiffness or endothelial dysfunction as a surrogate marker of vascular risk. Can J Cardiol 2006;22(suppl B):72B–80B.
10 Tawadrous H, Kamran H, Salciccioli L, et al: Evaluation of arterial structure and function in pediatric patients with end-stage renal disease on dialysis and after renal transplantation. Pediatr Transplant 2012;16:480–485.
11 Colì L, Ursino M, Donati G, Cianciolo G, et al: Clinical application of sodium profiling in the treatment of intradialytic hypotension. Int J Artif Organs 2003;26:715–722.
12 Colì L, Donati G, Cappuccilli ML, et al: Role of the hemodialysis vascular access type in inflammation status and monocyte activation. Int J Artif Organs 2011;34:481–488.
13 Roy-Chaudhury P, Spergel LM, Besarab A, et al: Biology of arteriovenous fistula failure. J Nephrol 2007;20:150–163.
14 Rothuizen TC, Wong C, Quax PH, et al: Arteriovenous access failure: more than just intimal hyperplasia? Nephrol Dial Transplant 2013;28:1085–1092.
15 Valente S, Pasquinelli G: Phenotypic and functional mapping of mesenchymal stem cells harvested from different portions of the human arterial tree; in Mesenchymal Stem Cell, ISBN 978-953-51-5280-4; InTech Rijeka 2017, in press.
16 Pasquinelli G, Tazzari PL, Vaselli C, et al: Thoracic aortas from multiorgan donors are suitable for obtaining resident angiogenic mesenchymal stromal cells. Stem Cells 2007; 25:1627–1634.
17 Pasquinelli G, Pacilli A, Alviano F, et al: Multidistrict human mesenchymal vascular cells: pluripotency and stemness characteristics. Cytotherapy 2010;12:275–287.
18 Hoshino A, Chiba H, Nagai K, et al: Human vascular adventitial fibroblasts contain mesenchymal stem/progenitor cells. Biochem Biophys Res Commun 2008;368:305–310.
19 Zengin E, Chalajour F, Gehling UM, et al: Vascular wall resident progenitor cells: a source for postnatal vasculogenesis. Development 2006;133:1543–1551.
20 Covas DT, Piccinato CE, Orellana MD, et al: Mesenchymal stem cells can be obtained from the human saphena vein. Exp Cell Res 2005;309:340–344.
21 Campagnolo P, Cesselli D, Al Haj Zen A, et al: Human adult vena saphena contains perivascular progenitor cells endowed with clonogenic and proangiogenic potential. Circulation 2010;121:1735–1745.
22 Biteau B, Hochmuth CE, Jasper H: Maintaining tissue homeostasis: dynamic control of somatic stem cell activity. Cell Stem Cell 2011;9:402–411.
23 Jager A, van Hinsbergh VW, Kostense PJ, et al: C-reactive protein and soluble vascular cell adhesion molecule-1 are associated with elevated urinary albumin excretion but do not explain its link with cardiovascular risk. Arterioscler Thromb Vasc Biol 2002;22:593–598.
24 Vaccaro F, Mulè G, Cottone S, et al: Circulating levels of adhesion molecules in chronic kidney disease correlate with the stage of renal disease and with C-reactive protein. Arch Med Res 2007;38:534–538.
25 Ciavarella C, Gallitto E, Ricci F, et al: The crosstalk between vascular MSCs and inflammatory mediators determines the pro-calcific remodelling of human atherosclerotic aneurysm. Stem Cell Res Ther 2017 (accepted).
26 Derwall M, Malhotra R, Lai CS, et al: Inhibition of bone morphogenetic protein signaling reduces vascular calcification and atherosclerosis. Arterioscler Thromb Vasc Biol 2012;32: 613–622.
27 Avogaro A, Fadini GP: Mechanisms of ectopic calcification: implications for diabetic vasculopathy. Cardiovasc Diagn Ther 2015;5: 343–352.

28 Orrico C, Pasquinelli G, Foroni L, et al: Dysfunctional vasa vasorum in diabetic peripheral artery obstructive disease with critical lower limb ischaemia. Eur J Vasc Endovasc Surg 2010;40:365–374.
29 Gostjeva EV, Zukerberg L, Chung D, Thilly WG: Bell-shaped nuclei dividing by symmetrical and asymmetrical nuclear fission have qualities of stem cells in human colonic embryogenesis and carcinogenesis. Cancer Genet Cytogenet 2006;164:16–24.
30 Fittipaldi S, Vasuri F, Degiovanni A, et al: The study of calcified atherosclerotic arteries: an alternative to evaluate the composition of a problematic tissue reveals new insight including metakaryotic cells. BMC Clin Pathol 2016;16:12.
31 Hayek SS, MacNamara J, Tahhan AS, et al: Circulating progenitor cells identify peripheral arterial disease in patients with coronary artery disease. Circ Res 2016;119:564–571.
32 Cianciolo G, Capelli I, Cappuccilli M, et al: Calcifying circulating cells: an uncharted area in the setting of vascular calcification in CKD patients. Clin Kidney J 2016;9:280–286.
33 Gostjeva EV, Koledova V, Tomita-Mitchell A, et al: Metakaryotic stem cell lineages in organogenesis of humans and other metazoans. Organogenesis 2009;5:191–200.
34 Gostjeva EV, Thilly WG: Stem cell stages and the origins of colon cancer: a multidisciplinary perspective. Stem Cell Rev 2005;1: 243–251.
35 Gruhl AN, Gostjeva EV, Thilly WG, et al: Human fetal/tumor metakaryotic stem cells: pangenomic homologous pairing and telomeric end-joining of chromatids. Cancer Genet Cytogenet 2010;203:203–208.
36 Thilly WG, Gostjeva EV, Koledova VV, et al: Metakaryotic stem cell nuclei use pangenomic dsRNA/DNA intermediates in genome replication and segregation. Organogenesis 2014;10:44–52.
37 Gostjeva EV, Koledova VV, Bai L, et al: Metakaryotic Cancer Stem Cells are Constitutively Resistant to X-Rays and Chemotherapeutic Agents, but Sensitive to Many Common Drugs; in Sherley JL Human Stem Cell Toxicology, London, Royal Society of Chemistry, 2016, pp 196–249.
38 Sudo H, Li-Sucholeiki XC, Marcelino LA, et al: Fetal-juvenile origins of point mutations in the adult human tracheal-bronchial epithelium: absence of detectable effects of age, gender or smoking status. Mutat Res 2008; 646:25–40.
39 Kini LG, Herrero-Jimenez P, Kamath T, et al: Mutator/hypermutable fetal/juvenile metakaryotic stem cells and human colorectal carcinogenesis. Front Oncol 2013;3:267.
40 Beaulieu JF, Ménard D: Isolation, characterization, and culture of normal human intestinal crypt and villus. Methods Mol Biol 2012; 806:157–173.
41 Garg A, Houlihan DD, Aldridge V, et al: Non-enzymatic dissociation of human mesenchymal stromal cells improves chemokine-dependent migration and maintains immunosuppressive function. Cytotherapy 2014; 16:545–559.
42 Chen F, Tillberg PW, Boyden ES: Expansion microscopy. Science 2015;347:543–547.

Prof. Gianandrea Pasquinelli, MD
S. Orsola-Malpighi Hospital
Bologna University
Via Massarenti 9, IT–40138 Bologna (Italy)
E-Mail gianandr.pasquinelli@unibo.it

La Manna G, Ronco C (eds): Current Perspectives in Kidney Diseases.
Contrib Nephrol. Basel, Karger, 2017, vol 190, pp 108–116 (DOI: 10.1159/000468955)

Mycophenolate Mofetil: A Possible Alternative Treatment for IgA Nephropathy

Olga Baraldi[a] · Giorgia Comai[a] · Vania Cuna[a] · Maria Cappuccilli[a] · Carla Serra[b] · Claudio Ronco[c, d] · Gaetano La Manna[a]

[a]Department of Experimental Diagnostic and Specialty Medicine (DIMES), Nephrology, Dialysis and Renal Transplant Unit, and [b]Department of Organ Failure and Transplantation/General Surgery, St. Orsola Hospital, University of Bologna, Bologna, [c]Department of Nephrology, Dialysis and Transplantation, and [d]International Renal Research Institute of Vicenza, San Bortolo Hospital, Vicenza, Italy

Abstract

IgA nephropathy (IgAN) is the most common primary glomerulonephritis in developed countries and a leading cause of chronic kidney disease. IgAN is a mesangial proliferative glomerulonephritis characterized by diffuse mesangial deposition of IgA, often accompanied by the deposition of IgG and the C3 component of complement in a similar distribution. This condition is in most cases oligosymptomatic, often discovered coincidentally. Currently, there is no specific treatment available for IgAN and the use of immunosuppression therapy is debated. Due to immune-mediated pathogenic nature of IgAN, therapy with mycophenolate mofetil (MMF), a potent immunosuppressive agent, could be effective in patients at risk for progressive disease. In this paper, we discuss the case of an IgAN patient treated with MMF at our center, followed by a review of the literature and our previous experience on the potential renoprotective effects of MMF in IgAN patients with different clinical presentation, despite adequate angiotensin blockade and steroid therapy.

Introduction

IgA nephropathy (IgAN) is the most common glomerulonephritis in developed countries. This condition is mostly an oligosymptomatic glomerulonephritis, characterized by episodic glomerular hematuria often with persistent proteinuria of a variable degree.

IgAN is regarded as an immune-mediated disease with the presence of circulating and glomerular immune complexes comprising of galactose-deficient IgA1, an IgG autoantibody directed against the hinge region O-glycans and C3.

The deposition of these under-galactosylated, dimeric or polymeric IgA antibodies in the glomerular mesangium results in a nephritogenic action contributing directly to glomerular inflammation and mesangial proliferation [1]. The presence of aberrantly glycosylated IgA1 is an heritable trait [2].

The diagnosis of IgAN requires histology evaluation of renal biopsy specimens and detection of IgA dominant or co-dominant deposits, frequently concomitant with mesangial/endocapillary hypercellularity and fibrosis, as outlined in the Oxford classification [3, 4].

The risk of disease progression is closely related to clinical parameters such as hypertension, proteinuria, and impaired glomerular filtration rate (GFR). Coexistent risk factors, such as older age, obesity, and smoking can contribute to disease progression through glomerular inflammation and microvascular injury [5, 6].

Established hypertension generally reflects a relatively advanced disease and is often related to a progressive course, if untreated. The presence of persistent proteinuria, although mild, has been repeatedly demonstrated to be associated with adverse prognosis, while partial remission of proteinuria to levels less than 1.0 g/day results in better outcomes [7].

Many patients have a benign disease, and often spontaneous remissions occur [8]. However, progressive renal failure is observed in about 30–50% of patients, necessitating renal replacement therapy after 25 years of follow-up.

Conservative treatment including blood pressure (BP) control, angiotensin converting enzyme inhibitors (ACEIs) or angiotensin receptor blockers (ARBs), lead to a reduction of proteinuria. However, there is still no specific treatment to date available for IgAN because its pathogenic mechanisms remain incompletely understood, and thus some patients not responsive to conventional treatments may experience a progression of renal disease. These subjects represent a " grey area," which is difficult to recognize and a more aggressive disease.

The Kidney Disease Improving Global Outcomes (KDIGO) guidelines regarding IgAN recommend treatment with a blocker of the RAAS in patients who have proteinuria with urinary protein excretion of more than 1 g/day. The KDIGO guidelines also suggest the use of systemic glucocorticoids in patients who have a proteinuria level above 1 g of urinary protein excretion per day and a GFR higher than 50 mL/min, despite supportive care [9].

As with other forms of glomerulonephritis, there are several trials supporting that corticosteroids reduce proteinuria in IgAN; however, the effectiveness of immunosuppressive drugs in patients who are at higher risk of progressive disease is unclear, and KDIGO Clinical Practice Guidelines have assigned low levels of evidence for almost all recommendations and suggestions related to this nephropathy [9].

Mycophenolate mofetil (MMF) might be an option for immunosuppressive treatment of patients with IgAN. Nevertheless, data regarding the efficacy and safety of MMF therapy in IgAN are controversial, because of the different patient profiles considered (low vs. high risk renal disease) in previous studies and small sample sizes [10].

In this paper, we describe a case report and our experience of IgAN patients treated with MMF, followed by a literature review for IgAN therapy.

Case Report

A 28-year-old male with biopsy proven IgAN with proteinuria 1 g/day was treated in 2002 with corticosteroids therapy in bolus (1 g intravenously for 3 days at months 1, 3, and 5, and oral prednisone at 0.5 mg/kg/day on alternate days for 6 months [11]) and ACEI.

Kidney function was always normal, and proteinuria decreased to 0.3 g/day after therapy.

In 2014, despite the ACEI therapy, proteinuria increased until 1 g/day, creatinine clearance dropped from 105 to 75 mL/min, BP and lipid profile were in the normal range. Therapy was modified with dual RAS blockade (ACEI plus ARB) without benefit. No hyperkalemia or hypotension occurred.

In 2015, a second kidney biopsy was performed and histology revealed a diffuse mesangial proliferative glomerulonephritis (Oxford classification: M0, E0, S1, T0). The patient started therapy with MMF 1.5 g/day, and 6 month later proteinuria was gradually reduced until the normal range and clearance creatinine returned to previous levels. Further laboratory examinations showed normal serum creatinine and hemoglobin levels.

Discussion

IgAN is a principal cause of end-stage renal disease (ESRD) worldwide in young adults. The mainstay of treatment for IgAN is the blockade of the RAAS. Agents able to reduce glomerular capillary hypertension can provide protection against glomerular injury. ACEI or ARB display direct effect on mesangial cell growth and proliferation; furthermore, inhibition of the RAAS has pleiotropic effects, including anti-inflammatory, anti-fibrotic, and anti-oxidant. The role of RAAS blockade had statistically significant effects on renoprotection and reduction of proteinuria [12]; combined treatment with ACEI plus ARB was not more effective than ACR or ARB alone [13].

There are several clinical trials supporting that corticosteroids reduce proteinuria in IgAN.

Pozzi et al. [11] reported that steroid treatment for 6 months (including 86 patients with proteinuria 1–3.5 g/day for at least 3 months and serum creatinine ≤1.5 g/dL) provided renoprotective effects, and the renal survival was better in the steroid group as compared with control after 10 years [14].

Similar results on the benefits of corticosteroids in proteinuric IgAN patients were demonstrated in 2 randomized controlled trials, one Italian and one Chinese, including respectively 97 and 63 patients with biopsy proven IgAN, proteinuria ≥1 g/day, and eGFR >50 mL/min/1.73 m^2 [15, 16].

Also, the KDIGO guidelines recommended that patients with persistent proteinuria >1 g/day, in spite of adequate ACEI or ARB, BP control, and GFR >50 mL/min/1.73 m^2, may receive a 6-month course of steroid therapy [9].

Therefore, the most effective treatment for IgAN patients with proteinuria and almost normal renal function appears to be a combination of RAAS blockers and steroids.

Recently, a retrospective analysis of a subgroup of patients of the VALIGA study, a large European cohort of 1,147 patients with IgAN, demonstrated improvements in terms of proteinuria and rate of renal function decline in patients receiving corticosteroid combined with ACEI or ARB, in contrast to RAAS blockade monotherapy even with an initial eGFR <50 mL/min/1.73 m^2 [17].

The use of the sequential cyclophosphamide/azathioprine regimen suggested by Ballardie and Roberts [18] (low-dose 1.5 mg/kg/day oral daily cyclophosphamide for 3 months followed by long-term low dose azathioprine plus low-dose steroids) in IgAN patients with significant decline in renal function, was associated with a marked protection from further decline and from ESRD.

In the recent STOP-IgAN trial (Supportive versus Immunosuppressive Therapy for the Treatment of Progressive IgAN, Rauen et al. [19] investigated the long-term benefits of a 6-month therapy of steroids in association with immunosuppression compared with conservative treatment in only patients who maintained proteinuria above 0.75 g/day after a comprehensive optimization of supportive treatment (including RAAS blockers for BP lowering to a target below 125/75 mm Hg, as well as lipid and smoking control). Of a total of 380 patients, only 162 with mild renal impairment (eGFR >30 mL/min/1.73 m^2) and persistent proteinuria received steroids and cyclophosphamide (1.5 mg/kg/day for 3 months), followed by maintenance immunosuppression with azathioprine (1.5 mg//kg/day), plus oral prednisolone at a dose of 40 mg/day gradually tapered. In this cohort, cyclophosphamide or high corticosteroid pulses did not demonstrate substantial improvements, considering both primary end-point ("full clinical remission," defined as proteinuria with a protein-to-creatinine

Table 1. Published controlled trials investigating effects of MMF in IgAN patients

Controlled trials	Sample size	Patient characteristics	Treatment	Follow-up	Efficacy
Maes et al. [25], 2004	34 (21 treatment arm, 13 control arm)	Severe IgAN, arterial hypertension, impaired renal function	MMF versus placebo	3 years	No
Frisch et al. [26], 2005	32 (17 treatment arm, 15 control arm)	Advance renal insufficiency	MMF versus placebo	2 years	No
Tang et al. [28], 2005	40 (20 treatment arm, 20 control arm)	Mild tubulointerstitial lesions, persistent proteinuria >1 g/day	MMF versus placebo	6 years	Yes, proteinuria reduction 62% from baseline. After 6 years, no difference in proteinuria but better renal function in MMF group
Roccatello et al. [30], 2012	8 patients	Florid glomerular changes, proteinuria >2.4 g/day, renal failure (mean serum creatinine 1.6 mg/dL)	Steroids 3 pulses followed by oral prednisone and MMF	2 years	Yes, serum creatinine and proteinuria dropped down versus baseline
Liu et al. [31], 2014	84 patients	Severe IgAN	MMF + steroids versus cyclophosphamide + steroids	18 months	Reduction of proteinuria and improvement in renal function in MMF group with less adverse events compared to Cyc group
Hogg et al. [27], 2015	44 (22 treatment arm, 22 control arm)	Urine protein-creatinine ratio >0.6 g/g	MMF versus placebo	6 months	No reduction in proteinuria
Hou et al. [32], 2017	176 patients	Active proliferative lesions, mean proteinuria 2.3 g/day	Steroids + MMF versus steroids	12 months	Complete remission in both groups, lower incidence of Cushing syndrome and diabetes mellitus in MMF group

ratio <0.2 and stable renal function with a decrease in the eGFR <5 mL/min/1.73 m^2 from the baseline eGFR) and secondary end-point, defined as a decrease in the eGFR of at least 15 mL/min/1.73 m^2 from the baseline eGFR. At 3 years, a large proportion of patients were in a full clinical remission compared with patients treated with supportive care. But these were not associated with a proven improvement in outcome. STOP-IgA trial is discussed controversially about patient selection, inclusion criteria, short follow-up period, and the lack of renal histology [19].

The underlying reason for unresponsiveness to conventional therapy observed in some patients is a topic of debate. MMF has been successfully used in the treatment of immunomediated glomerular disorders [20] and the pathogenesis of IgAN, although incompletely understood, is recognized as an autoimmune renal disease.

The effectiveness of MMF in patients with IgAN has been investigated in different randomized clinical trials with controversial results (Table 1).

MMF is activated by plasma and tissue esterases to the active mycophenolic acid (MPA). Although the genes coding for esterases are highly polymorphic and this might partly explain the interindividual differences in bioactivation, further investigations using human liver microsomes has revealed that allelic variants make only a minor contribution to MPA pharmacokinetic variability [21]. MPA is eliminated by conjugation in the liver to glucuronide metabolites that can then be reactivated by enterohepatic recycling to MPA, causing the typical secondary MPA peak in the concentration-time curve, usually between 6 and 8 h after a dose. Pharmacogenetic factors can influence enterohepatic recycling of MPA, as part of excretion and elimination, and provide an explanation for the variability in response [22–24]. MPA is metabolized to an inactive aryl glucuronide, important for enterohepatic recycling, and a pharmacologically active acyl glucuronide, through UDP-glucuronosyltransferase isoforms, mainly 1A8, 1A9, and 2B7 [23].

However findings from these studies have been contradictory because MMF was used with different outcomes, in relatively small cohorts of patients, and in some cases without steroid therapy. Maes et al. [25] investigated 34 patients with impaired renal function, histologically unfavorable criteria and arterial hypertension. All subjects were treated with salt intake restriction, ACEI, and MMF 2 g/day ($n = 21$) or placebo ($n = 13$), but the results failed to demonstrate a beneficial effect of MMF in terms of kidney outcome or proteinuria after 3 years of evaluation.

Other trials on patients with advanced renal insufficiency reported a worse outcome in MMF group, suggesting that MMF is probably not effective in IgAN patients with an already established moderate renal insufficiency [26, 27].

In contrast, in a small, randomized trial of 40 Chinese patients with persistent proteinuria >1 g/day, MMF therapy resulted in a stable remission of proteinuria, even 12 months after stopping therapy [28], and long-term follow-up demonstrated improved renal survival in MMF-treated patients [10].

An Italian multicenter study showed that in patients with active disease, proteinuria >2.4 g/day and renal failure (mean serum creatinine 1.6 mg/dL), a combined regimen with MMF and steroid can induce remission of proteinuria and slower successive progression toward renal failure [29].

In 2014, Liu et al. [30] compared combined therapy with prednisone and MMF versus prednisone and cyclophosphamide for severe IgAN. In MMF group, a reduction of proteinuria and improvement in renal function was observed, with less adverse events compared to the cyclophosphamide group.

Very recently, Hou et al. [31] evaluated the efficacy and safety of MMF plus prednisone in comparison to full-dose prednisone in patients with IgAN with active proliferative lesions. The primary end point was the rate of complete remission at 6 and 12 months, while the secondary end points included the rate of overall response (complete remission plus partial remission) at 6 and 12 months, median time to complete remission, relapse rate after stopping the treatments, and changes in active proliferative lesions on a repeat biopsy. At 6 months, there were no significant differences in both groups about complete remission (37 vs. 38%, $p = 0.9$) and median time to achieve remission (8.7 vs. 8.5 months, $p = 0.6$). During the follow-up period, relapse rates and total adverse events did not differ between the groups; repeat kidney biopsy data proved that endocapillary hypercellularity, crescents, and necrosis lesions in glomeruli had disappeared or improved after immunosuppressive therapy.

In our previous experience, we had 2 cases of patients with IgAN treated with MMF. A 27-year-old woman was referred to our Nephrology Unit due to the presence of microhematuria and proteinuria (1.5 g/day) at urine examination. The patient had normal renal function, mild hypertension, and was overweight. Kidney biopsy showed mesangial hypercellularity with focal segmental glomerulosclerosis with adhesions (M0, E0, S1, T0). She was treated with ACEI, and steroid therapy was given for a short period to avoid metabolic adverse effect in a young, overweight patient. Then a regimen based on MMF (1.5 g/day) for 24 months combined with ACEI was started with progressive improvement of proteinuria until complete remission (proteinuria 0.2 g/day). After 5 years of follow-up, complete remission was achieved and the current therapy consisted of ACEI.

A second case was a 54-year-old man who underwent kidney biopsy due to renal damage (serum creatinine 2 mg/dL) and proteinuria 3.5 g/day. Histological examination showed mesangial hypercellularity, endocapillary hypercellularity, focal segmental glomerulosclerosis, and crescents. The Oxford classification was M1, E1, S1, T0. He started steroid treatment combined with cyclophosphamide both in bolus followed by MMF (2 g/day). After 2 years, a reduction of proteinuria and improvement in renal function was observed.

Univocal treatment of IgAN is still largely based on opinion or weak evidence. Taken together, our cases reflect the wide spectrum of histologic lesions that characterized IgAN, ranging from stable disease with physiological decrease of renal function and low grade of proteinuria to progressive disease with relative rapid decline in renal function.

Nevertheless, several trials found that full-dose prednisone resulted in better clinical outcomes in terms of remission and showed proven benefits in both long-term use and pulse therapy [14, 15]. Corticosteroids display anti-inflammatory effects that could be responsible for a reduction of the proliferative lesions, glomerular sclerosis, and tubular fibrosis in IgAN.

MMF has been used in heterogeneous, relatively small cohorts of patients with various histologic lesions. Hence, it is difficult to compare in terms of outcomes or risk factors such as proteinuria, age or progression of disease. Ethnicity or pharmacogenetic aspects may be other possible reasons to account for the differences observed in these studies.

To date, it is not clear whether IgAN patients could have some advantages from "induction" therapy and "maintenance" therapy like in lupus nephritis or ANCA-associated glomerulonephritis.

Data from published studies clearly demonstrated that corticosteroids are necessary in combination with MMF for long term-maintenance. Moreover, MMF therapy resulted in a lower incidence of Cushing syndrome and newly diagnosed diabetes mellitus.

In conclusion, MMF therapy might represent a possible alternative to long-term steroid treatment or other immunosuppressive regimen such as azathioprine or cyclophosphamide in patients with persistent proteinuria or abnormal decline in renal function.

References

1 Rodrigues JC, Haas M, Reich HN: IgA nephropathy. Clin J Am Soc Nephrol 2017;pii:CJN.07420716.

2 Kiryluk K, Moldoveanu Z, Sanders JT, et al: Aberrant glycosylation of IgA1 is inherited in both pediatric IgA nephropathy and Henoch-Schonlein purpura nephritis. Kidney Int 2011;80:79–87.

3 Cattran DC, Coppo R, Cook HT, et al; Working Group of the International IgA Nephropathy Network and the Renal Pathology Society: The Oxford classification of IgA nephropathy: rationale, clinicopathological correlations, and classification. Kidney Int 2009;76:534–545.

4 Coppo R, Troyanov S, Camilla R, et al: The Oxford IgA nephropathy clinicopathological classification is valid for children as well as adults. Kidney Int 2010;77:921–927.

5 Myllymaki J, et al: Uric acid correlates with the severity of histopathological parameters in IgA nephropathy. Nephrol Dial Transplant 2005;20:89–95.

6 Yamamoto R, Nagasawa Y, Shoji T, et al: Cigarette smoking and progression of IgA nephropathy. Am J Kidney Dis 2010;56:313–324.

7 Reich HN, Troyanov S, Scholey JW, et al: Remission of proteinuria improves prognosis in IgA nephropathy. J Am Soc Nephrol 2007;18:3177–3183.

8 Gutierrez E, Zamora I, Ballarin JA, et al: Long-term outcomes of IgA nephropathy presenting with minimal or no proteinuria. J Am Soc Nephrol 2012;23:1753–1760.

9 KDIGO clinical practice guideline for glomerulonephritis. Kidney Int Suppl 2012;2:142.

10 Tang SC, Tang AW, Wong SS, et al: Long-term study of mycophenolate mofetil treatment in IgA nephropathy. Kidney Int 2010;77:543–549.

11 Pozzi C, Bolasco PG, Fogazzi GB, et al: Corticosteroids in IgA nephropathy: a randomised controlled trial. Lancet 1999;353:883–887.

12 Cheng J, Zhang X, Tian J, et al: Combination therapy an ACE inhibitor and an angiotensin receptor blocker for IgA nephropathy: a meta-analysis. Int J Clin Pract 2012;66:917–923.

13 Reid S, Cawthon PM, Craig JC, et al: Non-immunosuppressive treatment for IgA nephropathy. Cochrane Database Syst Rev 2011;3:CD003962.

14 Pozzi C, Andrulli S, Del Vecchio L, et al: Corticosteroid effectiveness in IgA nephropathy: long-term results of a randomized, controlled trial. J Am Soc Nephrol 2004;15:157–163.

15 Manno C, Torres DD, Rossini M, et al: Randomized controlled clinical trial of corticosteroids plus ACE-inhibitors with long-term follow-up in proteinuric IgA nephropathy. Nephrol Dial Transplant 2009;24:3694–3701.

16 Lv J, Zhang H, Chen Y, et al: Combination therapy of prednisone and ACE inhibitor versus ACE-inhibitor therapy alone in patients with IgA nephropathy: a randomized controlled trial. Am J Kidney Dis 2009;53: 26–32.
17 Tesar V, Troyanov S, Bellur S, et al: Corticosteroids in IgA nephropathy: a retrospective analysis from the VALIGA study. J Am Soc Nephrol 2015;26:2248–2258.
18 Ballardie FW, Roberts IS: Controlled prospective trial of prednisolone and cytotoxics in progressive IgA nephropathy. J Am Soc Nephrol 2002;13:142–148.
19 Rauen T, Eitner F, Fitzner C, et al; STOP-IgAN Investigators: Intensive supportive care plus immunosuppression in IgA nephropathy. N Engl J Med 2015;373:2225–2236.
20 Sepe V, Libetta C, Giuliano MG, et al: Mycophenolate mofetil in primary glomerulopathies. Kidney Int 2008;73:154–162.
21 Fujiyama N, Miura M, Satoh S, et al: Influence of carboxylesterase 2 genetic polymorphisms on mycophenolic acid pharmacokinetics in Japanese renal transplant recipients. Xenobiotica 2009;39:407–414.
22 Barraclough KA, Lee KJ, Staatz CE: Pharmacogenetic influences on mycophenolate therapy. Pharmacogenomics 2010;11:369–390.
23 Dupuis R, Yuen A, Innocenti F: The influence of UGT polymorphisms as biomarkers in solid organ transplantation. Clin Chim Acta 2012;413:1318–1325.
24 Miura M, Satoh S, Inoue K, et al: Influence of SLCO1B1, 1B3, 2B1 and ABCC2 genetic polymorphisms on mycophenolic acid pharmacokinetics in Japanese renal transplant recipients. Eur J Clin Pharmacol 2007;63: 1161–1169.
25 Maes BD, Oyen R, Claes K, et al: Mycophenolate mofetil in IgA nephropathy: results of a 3-year prospective placebo-controlled randomized study. Kidney Int 2004;65:1842–1849.
26 Frisch G, Lin J, Rosenstock J, et al: Mycophenolate mofetil (MMF) vs placebo in patients with moderately advanced IgA nephropathy: a double-blind randomized controlled trial. Nephrol Dial Transplant 2005;20:2139–2145.
27 Hogg RJ, Bay RC, Jennette JC, et al: Randomized controlled trial of mycophenolate mofetil in children, adolescents, and adults with IgA nephropathy. Am J Kidney Dis 2015;66:783–791.
28 Tang S, Leung JC, Chan LY, et al: Mycophenolate mofetil alleviates persistent proteinuria in IgA nephropathy. Kidney Int 2005;68: 802–812.
29 Roccatello D, Rossi D, Marletto F, et al: Long-term effects of methylprednisolone pulses and mycophenolate mofetil in IgA nephropathy patients at risk of progression. J Nephrol 2012;25:198–203.
30 Liu X, Dewei D, Sun S, et al: Treatment of severe IgA nephropathy: mycophenolate mofetil/prednisone compared to cyclophosphamide/prednisone. Int J Clin Pharmacol Ther 2014;52:95–102.
31 Hou JH, Le WB, Chen N, et al: Mycophenolate mofetil combined with prednisone versus full-dose prednisone in IgA nephropathy with active proliferative lesions: a randomized controlled trial. Am J Kidney Dis 2017;pi i:S0272-6386(17)30008-2.

Gaetano La Manna, MD, PhD
Department of Experimental, Diagnostic and Specialty Medicine (DIMES)
Nephrology, Dialysis and Transplantation Unit
St. Orsola Hospital, University of Bologna
Via G. Massarenti 9 (Pad. 15), IT–40138 Bologna (Italy)
E-Mail gaetano.lamanna@unibo.it

La Manna G, Ronco C (eds): Current Perspectives in Kidney Diseases.
Contrib Nephrol. Basel, Karger, 2017, vol 190, pp 117–123 (DOI: 10.1159/000468956)

Genetics and Autosomal Dominant Polycystic Kidney Disease Progression

Valentina Corradi[a, b] · Anna Giuliani[a] · Fiorella Gastaldon[a] · Massimo de Cal[a, b] · Barbara Mancini[c] · Anna Montaldi[c] · Alberta Alghisi[c] · Irene Capelli[d] · Gaetano La Manna[d] · Claudio Ronco[a, b]

[a]Department of Nephrology, Dialysis and Transplantation, [b]International Renal Research Institute (IRRIV), and [c]Genetics and Molecular Biology, Transfusion Medicine, San Bortolo Hospital, Vicenza, and [d]Department of Experimental Diagnostic and Specialty Medicine (DIMES), Nephrology, Dialysis and Renal Transplant Unit, St. Orsola Hospital, University of Bologna, Bologna, Italy

Abstract

Autosomal dominant polycystic kidney disease (ADPKD) is the most common hereditary kidney disease, accounting for 10% of European patients on renal replacement therapy. In the previous years, many approaches to slow the progression of ADPKD were studied and many clinical trials published. In addition to having diagnostic role, the description of the genotype is even important to predict the progression of the disease and contributes, combined with several other factors, to a more precise patients classification. With the availability of disease-modifying drugs, "fast progression factors" are needed to early recognize those patients who would likely progress, before cyst growth reaches a critical value. ERA-EDTA working group on inherited kidney disorders included a series of recommendations resulting in a hierarchical decision algorithm to select patients who are most likely to benefit from the treatment. Beyond diagnosis, we will also discuss the important role of genetics in ADPKD progression and management.

Autosomal dominant polycystic kidney disease (ADPKD) is the most common hereditary kidney disease, accounting for 10% of European patients on renal replacement therapy [1]. ADPKD is characterized by the development of cysts in both kidneys and progressive kidney function loss [2, 3]. Cysts growth and enlarged polycystic kidneys are the main features of ADPKD, which is progressively associated with the reduction of glomerular filtration rate (GFR). However, while cysts development occurs early in life, GFR remains stable until at least 50% of parenchyma has been destroyed, so that end-stage renal disease (ESRD) usually occurs by the end of the fifth decade of life.

However, ESRD onset has a wide variation, both intra- and inter-familial, and even if the median age at dialysis start is 58 years in 70% of patients, there are some patients reaching dialysis before 40 years of age and some others never requiring dialysis [4].

This wide clinical variability reflects the genetic heterogeneity of the disease. *PKD1* and *PKD2* mutations are responsible for the disease, while the type of mutations and several gene polymorphisms, together with environmental factors are responsible for the severity of the disease [5]. In a study of patients with known PKD: 85% of ADPKD cases are related to mutations in the *PKD1* gene (located on chromosome 16, at p13.3 level), while the remaining 15% are related to mutations in the *PKD2* gene (located on chromosome 4, at q21-q23 level) [6]. A high level of allelic heterogeneity is found for both genes: to date, a total of 1,273 different mutations are reported for *PKD1* and 202 for *PKD2* [7]. The age at disease onset, the severity, and the clinical outcome contribute to the clinical variability intra- and inter-familial.

In 1992, Ravine et al. [8] showed that non-*PKD1* patients had a much milder phenotype compared to *PKD1*-linked patients. The median survival of non-*PKD1* patients was 71.5 years (56.0 years in *PKD1* patients) and they had a lower risk of progressing to renal failure (OR 0.35, 95% CI 0.13–0.92) [9, 10].

In the previous years, many approaches to slow the progression of ADPKD were studied and many clinical trials published. In 2015, the European Medicines Agency (EMA) approved Tolvaptan (vasopressin V2 receptor antagonist) for ADPKD patients. This authorization was based on the results of TEMPO 3: 4 study, which was a randomized, placebo-controlled, 3-year trial, in which 1,445 patients with ADPKD aged 18–45 years, with total kidney volume (TKV) measured with magnetic resonance imaging (MRI) >750 mL and estimated creatinine clearance >60 mL/min, were randomly assigned to assume tolvaptan or placebo. Tolvaptan, the V2 receptor antagonist, which interferes with the intracellular signaling of cAMP, was shown to slow the rate of TKV growth by 49%, from 5.5 to 2.8% per year, and the rate of estimated GFR loss by 26%, from 3.7 to 2.72 mL/min/1.73 m^2 per year [11].

However, EMA authorized tolvaptan only for adult ADPKD patients with CKD stages 1–3 and with evidence of rapid progressive disease. This raised the question on how to identify patients with rapid progressive disease, who are eligible for treatment.

With the aim of giving treatment guidance in clinical practice, ERA-EDTA working group [12] on inherited kidney disorders suggested the use of an algorithm in which GFR and TKV are the 2 main elements to define the risk of rap-

Table 1. PROPKD score

Being male	1 point
Hypertension before 35 years of age	2 points
First urological event before 35 years of age	2 points
PKD2 mutation	0 point
Non-truncating PKD1 mutation	2 points
Truncating PKD1 mutation	4 points

id progression. They included a series of recommendations resulting in a hierarchical decision algorithm to select patients who are most likely to treatment benefit. They also suggested that in patients with a truncating mutation in the *PKD1* gene, in conjunction with early onset of clinical symptoms, consistent with a PROPKD score of >6, rapid progression is likely. The "PROPKD" risk scoring system [13], proposed by the Genkyst cohort study, aims to better predict the evolution of the ADPKD disease by combining clinical and genetic data. It is based on 4 variables: sex (male, 1 point), hypertension, first urologic event before 35 years of age (both 2 points), and PKD mutation. The highest prognostic weighting was associated with the genotype: 4 points for truncating *PKD1* mutations, 2 points for non-truncating *PKD1* mutation, and 0 points for *PKD2* mutations. The PRO-PKD score was calculated as the sum of 4 variable weightings with results ranging from 0 to 9 points (Table 1). The 3 risk categories were defined as: low risk (0–3 points), intermediate (4–6), and high risk (7–9) of progression to ESRD.

Nonsense mutations occur when a premature nonsense or stop codon is introduced in the DNA sequence. When the mutated sequence is translated into a protein, the resulting protein is incomplete and shorter than normal. Consequently, most nonsense mutations result in non-functional proteins [14].

In the Genkyst cohort study, Cornec-Le Gall et al. [15] showed a strong effect of *PKD1* mutation type on renal phenotype: 12-years delay in ESRD onset in patients with *PKD1* non-truncating mutation (i.e., in-frame mutation and missense events) compared to *PKD1* patients harboring truncating mutation (i.e., frameshift, nonsense, splice mutations, and large rearrangements; 95% CI 62.4–73.4 vs. 95% CI 53.6–57.7 years, $p < 0.0001$). The risk of developing ESRD was 2.74 times higher in patients with a truncating mutation than in patients with a non-truncating mutation of *PKD1*.

Also, in the 2000s, Rossetti et al. [16, 17] demonstrated that the position of the *PKD1* mutation correlates with the severity of renal disease. Patients with mutations in the 5′ region had significantly more severe disease than the 3′ group; the median time to ESRD was 53 and 56 years, respectively ($p = 0.025$), with less than half the chance of adequate renal function at 60 years (18.9 and 39.7%, respectively). Furthermore, they describe a number of unusual families with atypical presentations of PKD that suggest a role for incompletely penetrant *PKD1* alleles in causing and modulating cystic disease. Homozygosity for *PKD1* and coinheritance in trans of an incompletely penetrant *PKD1* allele with an inactivating *PKD1* mutation may explain some of the cases of early-onset PKD [18], suggesting that the dosage of the polycystin1 (*PKD1* functional protein) can contribute to the phenotypic variability [19]. Recently, the study of Ali et al. [20] proposes that the novel variant aggravated the disease phenotype, resulting in the early onset of ESRD and renal enlargement.

Moreover, high complexity of phenotypic variability may be caused by other modifier genes [21]. The extreme intra-familial variability in the phenotypic expression of *PKD1* is consistent with the hypothesis that one or a few modifier loci have a major effect on the progression and severity of *PKD1* [22]. Several gene polymorphisms have been associated with renal outcome, but none of the polymorphisms studied plays a major role in disease progression. Hypomorphic alleles and mosaicism may also account for some cases with unusually mild disease [23, 24], as well as early and severe phenotypes of ADPKD are attributable to additional PKD genes [25].

Moreover, 7–10% of ADPKD-affected patients are genetically unresolved (GUR), suggesting further genetic heterogeneity of this disorder. Recently, whole-exome sequencing of GUR ADPKD-affected families identified a missense mutation in GANAB, a gene encoding glucosidase II subunit a (GIIa). Analysis of GANAB-null cells showed an absolute requirement of GIIa for maturation and surface and ciliary localization of ADPKD proteins (PC1 and PC2) [26]. This mutation was associated with a mild phenotype. Difficulties in genetic investigations, and the presence of non-PKD1 and non-PKD2-related mutations, can make the predictability of renal function on a genetic basis more complex.

The diagnosis of ADPKD is established principally by renal imaging modalities [27], described by unified criteria for ultrasonographic diagnosis of ADPKD [28]. Because of cost and safety, ultrasonography is most commonly used as the standard imaging modality. In certain settings, genetic testing is required for a definitive diagnosis, particularly in young individuals. Consequently, it plays an increasingly important role in the diagnosis and management of ADPKD patients. As described before, of all the prognostic genetic factors, the type of mutation (especially truncating mutation) represents the best indicator of the progression of ADPKD. Nevertheless, genetic testing is reserved for a small number of patients.

The advent of next-generation DNA sequencing (NGS) technologies has revolutionized the field of human genetics and molecular diagnostics [29, 30], reducing the cost and working time significantly. The NGS approach presented by Tan et al. [31] or Eisenberger et al. [32] is a highly accurate and reliable approach for mutation analysis by NGS, despite the complexity of interpretation of these variants.

To date, the main indications for genetic test are renal ultrasound atypical ADPKD with unusual clinical manifestations, ADPKD "de novo" without any family history, potential living, related donor, prenatal and pre-implantation genetic diagnosis, screening for the diagnosis of ADPKD in an asymptomatic individual at risk because of a positive family history [33] and when the imaging diagnosis is unclear [34].

In addition to having diagnostic role, the description of the genotype is even important to predict the progression of the disease and contributes, combined with several other factors, to a more precise patients classification. A scenario that becomes dangerous if not controlled. Therefore, the role of genetics in the pre-test phase and the assessment and understanding of the results are important.

Now that disease-modifying drugs are available, "fast progression factors" are needed to early recognize those patients who are likely to progress, before cyst growth reaches a critical value. Given the fact that ADPKD is a genetic disease and genetic mutation is mostly responsible for the severity of the disease, genetic factors are the best candidates for the early prediction of rapid progressive disease. Implementation of genetic analysis would increase not only our knowledge in the field of ADPKD, but would also be helpful in the early identification of such patients who would likely progress, and probably would benefit the most of therapy.

References

1 Spithoven EM, Kramer A, Meijer E, Orskov B, Wanner C, Abad JM, Areste N, de la Torre RA, Caskey F, Couchoud C, Finne P, Heaf J, Hoitsma A, de Meester J, Pascual J, Postorino M, Ravani P, Zurriaga O, Jager KJ, Gansevoort RT: Renal replacement therapy for autosomal dominant polycystic kidney disease (ADPKD) in Europe: prevalence and survival – an analysis of data from the ERA-EDTA registry. Nephrol Dial Transplant 2014;29(suppl 4):iv15–iv25.

2 Gabow PA: Autosomal dominant polycystic kidney disease. N Engl J Med 1993;329:332–342.

3 Torres VE, Harris PC: Autosomal dominant polycystic kidney disease: the last 3 years. Kidney Int 2009;76:149–168.

4 Spithoven EM, Kramer A, Meijer E, Orskov B, Wanner C, Caskey F, Collart F, Finne P, Fogarty DG, Groothoff JW, Hoitsma A, Nogier MB, Postorino M, Ravani P, Zurriaga O, Jager KJ, Gansevoort RT: Analysis of data from the ERA-EDTA registry indicates that conventional treatments for chronic kidney disease do not reduce the need for renal replacement therapy in autosomal dominant polycystic kidney disease. Kidney Int 2014; 86:1244–1252.

5 Corradi V, Gastaldon F, Caprara C, Giuliani A, Martino F, Ferrari F, Ronco C: Predictors of rapid disease progression in autosomal dominant polycystic kidney disease. Minerva Med 2017;108:43–56.
6 Rossetti S, Consugar MB, Chapman AB, Torres VE, Guay-Woodford LM, Grantham JJ, Bennett WM, Meyers CM, Walker DL, Bae K, Zhang QJ, Thompson PA, Miller JP, Harris PC: Comprehensive molecular diagnostics in autosomal dominant polycystic kidney disease. J Am Soc Nephrol 2007;18:2143–2160.
7 Autosomal dominant polycystic kidney disease mutation database (PKDB). http://pkdb-mayoedu/.
8 Ravine D, Walker RG, Gibson RN, Forrest SM, Richards RI, Friend K, Sheffield LJ, Kincaid-Smith P, Danks DM: Phenotype and genotype heterogeneity in autosomal dominant polycystic kidney disease. Lancet 1992; 340:1330–1333.
9 Hateboer N, v Dijk MA, Bogdanova N, Coto E, Saggar-Malik AK, San Millan JL, Torra R, Breuning M, Ravine D: Comparison of phenotypes of polycystic kidney disease types 1 and 2. European PKD1-PKD2 Study Group. Lancet 1999;353:103–107.
10 Hateboer N, Veldhuisen B, Peters D, Breuning MH, San-Millan JL, Bogdanova N, Coto E, van Dijk MA, Afzal AR, Jeffery S, Saggar-Malik AK, Torra R, Dimitrakov D, Martinez I, de Castro SS, Krawczak M, Ravine D: Location of mutations within the PKD2 gene influences clinical outcome. Kidney Int 2000; 57:1444–1451.
11 Torres VE, Chapman AB, Devuyst O, Gansevoort RT, Grantham JJ, Higashihara E, Perrone RD, Krasa HB, Ouyang J, Czerwiec FS: Tolvaptan in patients with autosomal dominant polycystic kidney disease. N Engl J Med 2012;367:2407–2418.
12 Gansevoort RT, Arici M, Benzing T, Birn H, Capasso G, Covic A, Devuyst O, Drechsler C, Eckardt KU, Emma F, Knebelmann B, Le Meur Y, Massy ZA, Ong AC, Ortiz A, Schaefer F, Torra R, Vanholder R, Wiecek A, Zoccali C, Van Biesen W: Recommendations for the use of tolvaptan in autosomal dominant polycystic kidney disease: a position statement on behalf of the ERA-EDTA working groups on inherited kidney disorders and European renal best practice. Nephrol Dial Transplant 2016;31:337–348.
13 Cornec-Le Gall E, Audrezet MP, Rousseau A, Hourmant M, Renaudineau E, Charasse C, Morin MP, Moal MC, Dantal J, Wehbe B, Perrichot R, Frouget T, Vigneau C, Potier J, Jousset P, Guillodo MP, Siohan P, Terki N, Sawadogo T, Legrand D, Menoyo-Calonge V, Benarbia S, Besnier D, Longuet H, Ferec C, Le Meur Y: The PROPKD score: a new algorithm to predict renal survival in autosomal dominant polycystic kidney disease. J Am Soc Nephrol 2016;27:942–951.
14 Lewis R: Human Genetics: Concepts and Applications. 9th edition, 2010.
15 Cornec-Le Gall E, Audrezet MP, Chen JM, Hourmant M, Morin MP, Perrichot R, Charasse C, Whebe B, Renaudineau E, Jousset P, Guillodo MP, Grall-Jezequel A, Saliou P, Ferec C, Le Meur Y: Type of PKD1 mutation influences renal outcome in ADPKD. J Am Soc Nephrol 2013;24:1006–1013.
16 Rossetti S, Burton S, Strmecki L, Pond GR, San Millan JL, Zerres K, Barratt TM, Ozen S, Torres VE, Bergstralh EJ, Winearls CG, Harris PC: The position of the polycystic kidney disease 1 (PKD1) gene mutation correlates with the severity of renal disease. J Am Soc Nephrol 2002;13:1230–1237.
17 Rossetti S, Harris PC: Genotype-phenotype correlations in autosomal dominant and autosomal recessive polycystic kidney disease. J Am Soc Nephrol 2007;18:1374–1380.
18 Rossetti S, Kubly VJ, Consugar MB, Hopp K, Roy S, Horsley SW, Chauveau D, Rees L, Barratt TM, van't Hoff WG, Niaudet P, Torres VE, Harris PC: Incompletely penetrant PKD1 alleles suggest a role for gene dosage in cyst initiation in polycystic kidney disease. Kidney Int 2009;75:848–855.
19 Bergmann C, von Bothmer J, Ortiz Bruchle N, Venghaus A, Frank V, Fehrenbach H, Hampel T, Pape L, Buske A, Jonsson J, Sarioglu N, Santos A, Ferreira JC, Becker JU, Cremer R, Hoefele J, Benz MR, Weber LT, Buettner R, Zerres K: Mutations in multiple PKD genes may explain early and severe polycystic kidney disease. J Am Soc Nephrol 2011;22:2047–2056.
20 Ali H, Hussain N, Naim M, Zayed M, Al-Mulla F, Kehinde EO, Seaburg LM, Sundsbak JL, Harris PC: A novel PKD1 variant demonstrates a disease-modifying role in trans with a truncating PKD1 mutation in patients with autosomal dominant polycystic kidney disease. BMC Nephrol 2015;16:26.

21 Pei Y: Molecular genetics of autosomal dominant polycystic kidney disease. Clin Invest Med 2003;26:252–258.
22 Fain PR, McFann KK, Taylor MR, Tison M, Johnson AM, Reed B, Schrier RW: Modifier genes play a significant role in the phenotypic expression of PKD1. Kidney Int 2005;67: 1256–1267.
23 Harris PC, Rossetti S: Determinants of renal disease variability in ADPKD. Adv Chronic Kidney Dis 2010;17:131–139.
24 Chang MY, Chen HM, Jenq CC, Lee SY, Chen YM, Tian YC, Chen YC, Hung CC, Fang JT, Yang CW, Wu-Chou YH: Novel PKD1 and PKD2 mutations in Taiwanese patients with autosomal dominant polycystic kidney disease. J Hum Genet 2013;58:720–727.
25 Bergmann C: ARPKD and early manifestations of ADPKD: the original polycystic kidney disease and phenocopies. Pediatr Nephrol 2015;30:15–30.
26 Porath B, Gainullin VG, Cornec-Le Gall E, Dillinger EK, Heyer CM, Hopp K, Edwards ME, Madsen CD, Mauritz SR, Banks CJ, Baheti S, Reddy B, Herrero JI, Bañales JM, Hogan MC, Tasic V, Watnick TJ, Chapman AB, Vigneau C, Lavainne F, Audrezet MP, Ferec C, Le Meur Y, Torres VE; Genkyst Study Group, HALT Progression of Polycystic Kidney Disease Group; Consortium for Radiologic Imaging Studies of Polycystic Kidney Disease, Harris PC: Mutations in GANAB, encoding the glucosidase IIα subunit, cause autosomal-dominant polycystic kidney and liver disease. Am J Hum Genet 2016;98: 1193–1207.
27 Torres VE, Harris PC, Pirson Y: Autosomal dominant polycystic kidney disease. Lancet 2007;369:1287–1301.
28 Pei Y, Obaji J, Dupuis A, Paterson AD, Magistroni R, Dicks E, Parfrey P, Cramer B, Coto E, Torra R, San Millan JL, Gibson R, Breuning M, Peters D, Ravine D: Unified criteria for ultrasonographic diagnosis of ADPKD. J Am Soc Nephrol 2009;20:205–212.
29 Kuhlenbaumer G, Hullmann J, Appenzeller S: Novel genomic techniques open new avenues in the analysis of monogenic disorders. Hum Mutat 2011;32:144–151.
30 Metzker ML: Sequencing technologies – the next generation. Nat Rev Genet 2010;11:31–46.
31 Tan AY, Michaeel A, Liu G, Elemento O, Blumenfeld J, Donahue S, Parker T, Levine D, Rennert H: Molecular diagnosis of autosomal dominant polycystic kidney disease using next-generation sequencing. J Mol Diagn 2014;16:216–228.
32 Eisenberger T, Decker C, Hiersche M, Hamann RC, Decker E, Neuber S, Frank V, Bolz HJ, Fehrenbach H, Pape L, Toenshoff B, Mache C, Latta K, Bergmann C: An efficient and comprehensive strategy for genetic diagnostics of polycystic kidney disease. PLoS One 2015;10:e0116680.
33 Izzi C, Liut F, Dallera N, Mazza C, Magistroni R, Savoldi G, Scolari F: [Genetics and genetic counseling]. G Ital Nefrol 2016; 33:pii:gin/33.2.4.
34 Scolari F, Savoldi G, Mazza C, Izzi C: [Molecular diagnosis of ADPKD]. G Ital Nefrol 2016;33:pii:gin/33.5.16.

Valentina Corradi
Department of Nephrology, Dialysis and Transplantation
San Bortolo Hospital, Viale Rodolfi, 37
IT–36100 Vicenza (Italy)
E-Mail valentina.corradi@aulss8.veneto.it

La Manna G, Ronco C (eds): Current Perspectives in Kidney Diseases.
Contrib Nephrol. Basel, Karger, 2017, vol 190, pp 124–133 (DOI: 10.1159/000468959)

Expanded Hemodialysis: A New Therapy for a New Class of Membranes

Claudio Ronco[a, b] · Gaetano La Manna[c]

[a]Department of Nephrology, Dialysis and Transplantation, [b]International Renal Research Institute of Vicenza, San Bortolo Hospital, Vicenza, and [c]Department of Experimental, Diagnostic and Specialty Medicine, Nephrology Dialysis and Transplantation Unit, St. Orsola Hospital, University of Bologna, Bologna, Italy

Abstract

A wide spectrum of molecules is retained in end-stage kidney disease, normally defined as uremic toxins. These solutes have different molecular weights and radii. Current dialysis membranes and techniques only remove solutes in the range of 50–15,000 Da, with limited or no capability to remove solutes in the middle to high molecular weight range (up to 50,000 Da). Improved removal has been obtained with high cut-off (HCO) membranes, with albumin loss representing a limitation to their practical application. Hemodiafiltration (HDF) at high volumes (>23 L/session) has produced some results on middle molecules and clinical outcomes, although complex hardware and high blood flows are required. A new class of membrane has been recently developed with a cut off (MWCO) close to the molecular weight of albumin. While presenting negligible albumin loss, these membranes have a very high retention onset (MWRO), allowing high clearances of solutes in a wide spectrum of molecular weights. These membranes originally defined (medium cut off) are probably better classified as high retention onset. The introduction of such membranes in the clinical routine has allowed the development of a new concept therapy called "expanded hemodialysis" (HDx). The new therapy is based on a special hollow fiber and dialyzer design. Its simple set-up and application offer the possibility to use it even in patients with suboptimal vascular access or even with an indwelling catheter. The system does not require a particular hardware or unusual nursing skill. The quality of dialysis fluid is, however, mandatory to ensure a safe conduction of the dialysis session. This new therapy is likely to modify the outcome of end-stage kidney disease patients, thanks to the enhanced removal of molecules traditionally retained by current dialysis techniques.

Introduction

In uremia, solutes over a wide molecular weight range display increased blood levels due to impaired kidney cleansing capacity [1]. While toxicity of small solutes has been known for a long time, the effect of retention of larger solutes and low molecular weight proteins (size range 5–35 kD) has recently emerged [2]. Among others, serum concentration of beta2-microglobulin and inflammatory mediators have been correlated with malnutrition-inflammation-atherosclerosis and formation of amyloid deposits in bone, tendons, and joints [3–5]. Other low molecular weight proteins such as free light chains, are retained or modified (glycosylation, oxidation) in uremia causing unwanted effects such as inflammation and cardiovascular complication [6]. Identification of products such as uremic toxins has promoted interest in increasing the removal of larger solutes. Originally, membranes for hemodialysis were designed to remove small solutes such as urea and creatinine while avoiding albumin loss [7]. The identification of toxins in the weight range of 500–5,000 Da confirmed the middle molecule hypothesis and spurred new interest in the development of new synthetic high flux membranes and their use in alternative techniques such as hemofiltration or hemodiafiltration (HDF) [8]. While chronic hemofiltration has been used only sporadically and presents the drawback of limited efficiency for small solute removal, HDF has progressively gained consensus, especially in Europe and has shown significant beneficial effects in clinical trials [9].

Nevertheless, even the more advanced dialysis techniques are still presenting high rates of cardiovascular complications and mortality [10].

Also, HDF requires a well-functioning vascular access, remarkably high blood flows, and a certain complexity of the hardware to achieve large amounts of convective transport. Thus, high flux dialysis still represents the most commonly used technique and is a compromise between simplicity and efficiency. More recently, new membranes with increased average diameter of pores and defined as "high cut-off (HCO)" have been used to remove myoglobin in trauma patients, albumin-bound toxins, inflammatory molecules in sepsis, and free light chains in hematological disorders, allowing to further expand the spectrum of molecules cleared [11]. The price of this membrane modification, however, is the increased number of large pores leading to some loss of albumin that is considered potentially harmful. Low serum albumin is a predictor of mortality in hemodialysis patients [12], and albumin loss through high-flux membranes (HF) processed for reuse with bleach is associated with a decrease in serum albumin concentration [11]. From a different point of view, a modest loss of albumin might even be considered beneficial as in the case of peritoneal dialysis patients [11]. While the question about albumin loss is still unanswered, this limits the application of these membranes in chronic hemodialysis, restricting their utilization to sporadic cases and in diffusive mode only. Based on these considerations, further steps are required to optimize large solute removal while retaining albumin as much as possible. These steps move in different directions including modification of polymer blending, surface modification and functionalization, modification of pore structure, size and distribution in a new class of membranes defined "high retention onset" (HRO) [13, 14].

Why a New Class of Membranes?

The low water permeability of original cellulosic membranes was considered an advantage in the absence of dialysis equipment capable of controlling water removal. The advent of ultrafiltration control systems led to the development and use of HF membranes that allowed improved middle molecule removal including β-2 microglobulin (β-2 M). Further advances in technology allowed better control over the structure and permeability of membranes. Different polymers and improved spinning modalities led to significant advances in solute removal and hemocompatibility. Inner surface modification produced a reduction in membrane thrombogenicity and protein-membrane interaction, with less inclination towards fouling and permeability decay. Further evolution in technology led to the development of a new class of membranes, referred to as protein-leaking membranes or super-flux or high cut off (HCO) [11]. These membranes are more permeable than conventional HF membranes and allow passage of larger proteins, including albumin. The rationale for these membranes is the need for increased clearance of low molecular weight proteins and protein-bound solutes. Protein-leaking membranes have been fabricated from a variety of polymers, including polymethylmethacrylate, cellulose triacetate, polysulfone, polyarylethersulfone, and polyethersulfone. The last evolution in the field of membranes is the development of a new class, previously called medium cut off and now defined as high retention onset (HRO) [15, 16]. Despite the definition, a more complex set of parameters characterize these membranes. To clarify the concept, we will use the schematic example reported in Figure 1, where 2 different membranes are compared. The typical sieving curve of membrane 1 describes the progressive reduction of the sieving value for a progressive increase of solute molecular weight, until the point at which 90% of the solute is retained in the filtration process. At this molecular weight, we define the cut-off value of the membrane (MWCO). The value of MWCO is similar for the 2 membranes, leading to the conclusion that these 2 membranes are similar. However, analyzing the molecular weight at which 10% of the solute begins to be retained, we can define the retention onset of the membrane (MWRO) [15, 16].

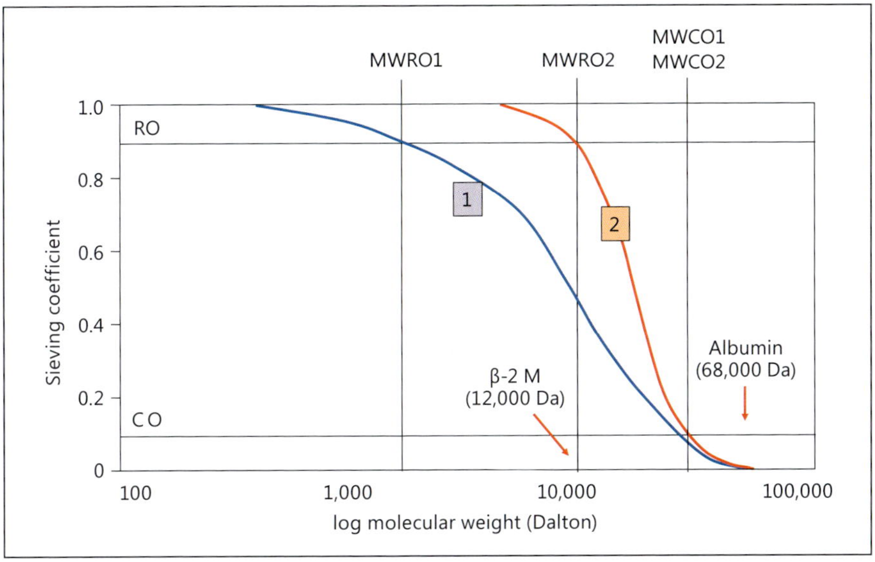

Fig. 1. Schematic representation of the sieving coefficient curves for 2 different classes of membranes: (1) high flux (HF) and (2) high retention onset (HRO). The point in the curve where the sieving coefficient is 0.1 determines the molecular weight cut-off (MWCO) value. The point in the curve where the sieving coefficient is 0.9 determines the molecular weight retention onset (MWRO) value. It is evident that HRO membrane, although presenting a similar cut-off value of the HF membrane, displays a completely different behavior. While MWRO for the HF membrane is in the range of 1,200 Da (vitamin B12), MWRO for the HRO membrane is in the range of 12,000 Da (β-2 microglobulin). MWCO value can be different in the 2 membranes with small leakage of albumin in the HRO class, but this effect is rapidly neutralized after few minutes from the beginning of the treatment due to protein deposition at the blood-membrane interface.

Based on the MWRO value, one can now differentiate the 2 membranes, which present very different properties and performance. Further speculation allows observing that membrane 2 has a MWRO in the MW range of β-2 microglobulin, while for that solute the sieving value is already reduced to 0.4 in the other membrane. This is what occurs with good approximation between HRO and HF membranes. In figure 2 we schematically describe the evolution of membrane development according to specific parameters.

The new class of membranes, therefore, has been developed to improve clearances of medium-large molecular weight solutes while avoiding albumin loss, as observed in the case of HCO membranes. This would partially contribute to answer those clinical needs that are relevant to medium-high molecular weight solute retention in uremia and its clinical consequences. In figure 3, we report the effect of sterical configuration of the molecule on its behavior in diffusion and convection.

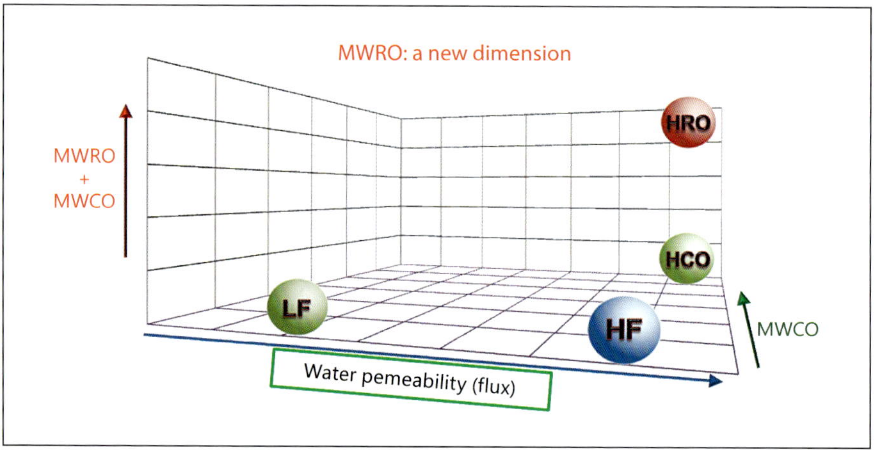

Fig. 2. The 3-dimensional graph describes the domain map of hemodialysis membranes. Three main parameters describe the nature and performance of the membrane: (1) water permeability or flux (increasing this parameter, we moved from low to high flux membranes), (2) membrane cut-off (MWCO), and (3) membrane retention onset (MWRO). The last 2 factors characterize the steepness of the sieving coefficient curve and its location in terms of molecular weight range.

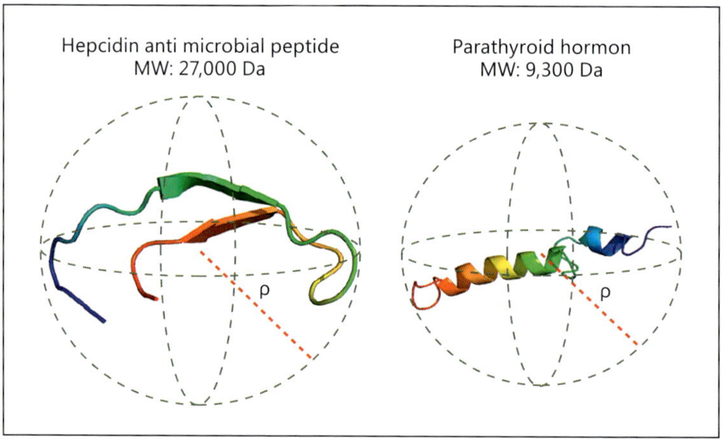

Fig. 3. There is a clear steric effect of the configuration of the molecules that make molecular weight just one of the elements affecting molecular behavior. In most cases, small proteins tend to fold and form irregular shapes. For this reason, it is better to consider the virtual molecular radius (Einstein-Stokes radius) or the radius of the sphere circumscribing the molecule. A schematic example is reported for hepcidin and human parathyroid hormone. The result is a diffusion coefficient and a sieving value lower than expected based on the effective molecular weight.

SOLUTE	MW (Da)	Class	Action/effect
Urea Creatinine Vitamin B12	60 125 1,250	Small	General toxicity
β-2 M Leptin Myoglobin	12,000 16,000 17,000	Middle	Amiloidosis CTS Malnutrition Organ damage
κ-FLC Prolactin Interleukin-6 Hepcidin BoundP-Cresol Pentraxin-3 λ-FLC TNF-α (Trim)	23,000 23,000 25,000 27,000 33,500 43,000 45,000 51,000	Large	Toxicity Infertility Inflammation Anemia CV toxicity Acute phase protein CV toxicity Inflammation
Albumin	68,000	Essential protein	Toxin binding capacity

Fig. 4. Example of some uremic toxins in a wide molecular weight spectrum. Standard HF membranes only clear few of them in the low and middle molecular weight ranges.

What Are the Target Retention Molecules?

Solute retention is a direct consequence of kidney failure. Different dialysis techniques are capable of keeping end-stage kidney disease patients alive and continue a nearly normal life. In the past, urea was considered the main target for renal replacement therapy; however, the real story of uremia is unfolding and several retention molecules are today held responsible for uremic complications [1–6]. In particular, retention of molecules in the middle-to-high molecular weight range cause inflammation, mineral bone disease, cardiovascular alterations, and accelerated atherosclerosis affecting long-term survival and patient's health. In Figure 4, there is a schematic summary of the uremic toxin pattern with specific reference to molecular weight of different retained molecules. Current high flux membranes are only effective in removing small and some middle molecules. All other molecules are retained, displaying a manifest inadequacy of current methods of renal replacement.

"Expanded Hemodialysis": A Way to Optimize the Use of HRO Membranes

A step forward has been taken with the use of high flux membranes in treatments with high convective components such as HDF. Due to the low diffusion coefficient of middle and large molecules, their removal in diffusive modalities is limited while the additional contribution of convection definitely improves the clearance pattern. In recent studies, HDF displayed significant advantages in terms of survival and comorbidities over standard high flux dialysis. Because convective clearance (K) results from the product of ultrafiltration rate (*Qf*) and sieving (*S*) of the selected molecule ($K = Qf \times S$), when the sieving is low, the only way to increase *K* is to increase *Qf*. In the past, this was difficult due to limited hardware capabilities, membrane fouling, low routine blood flows, and requirement of expensive bags of substitution fluid. With the advent of on-line HDF, the problem of fluid procurement has been solved and high convection rates have been made possible, thanks to the combined pre- and post-dilution configuration [9]. Nevertheless, on-line HDF is not approved in many countries including United States and still requires a complex hardware with multiple step filtration of incoming dialysate to ensure maximum purity of the reinfusion fluid. How can HRO membranes improve the efficacy of renal replacement therapy, and above all, how should they be utilized to exploit their new characteristics? We believe that the best application for HRO membranes is a new therapy called "expanded hemodialysis" (HDx). In Figure 5, we report the main points of this new therapy and the conditions required to exploit membrane capacity at best. HRO membrane must operate in hollow fiber configuration with a reduced inner diameter of the fiber. Moving from the standard 200 microns to 180 may allow increasing the wall shear rate and blood velocity per single fiber. This results in less fouling at the blood-membrane interface and improved solute exchange. A further effect is an increased end-to-end pressure drop with significant implications on the cross-filtration profile along the length of the fiber [17]. The fiber bundle should have adequate number of fibers to reach a surface area of at least 1.6 m^2. If the number of fibers is crucial to determine the cross-sectional area of the dialyzer, the length of the fibers and thus of the dialyzer will be quintessential to optimize internal filtration and the mechanism of filtration-backfiltration. This mechanism, although invisible, allows significant amount of convection inside the dialyzer where filtration takes place in the proximal part and backfiltration compensates in the distal part. The ultrafiltration control system of the dialysis machine regulates the process and provides the exact amount of net filtration required for the scheduled weight loss of the patient [18–20].

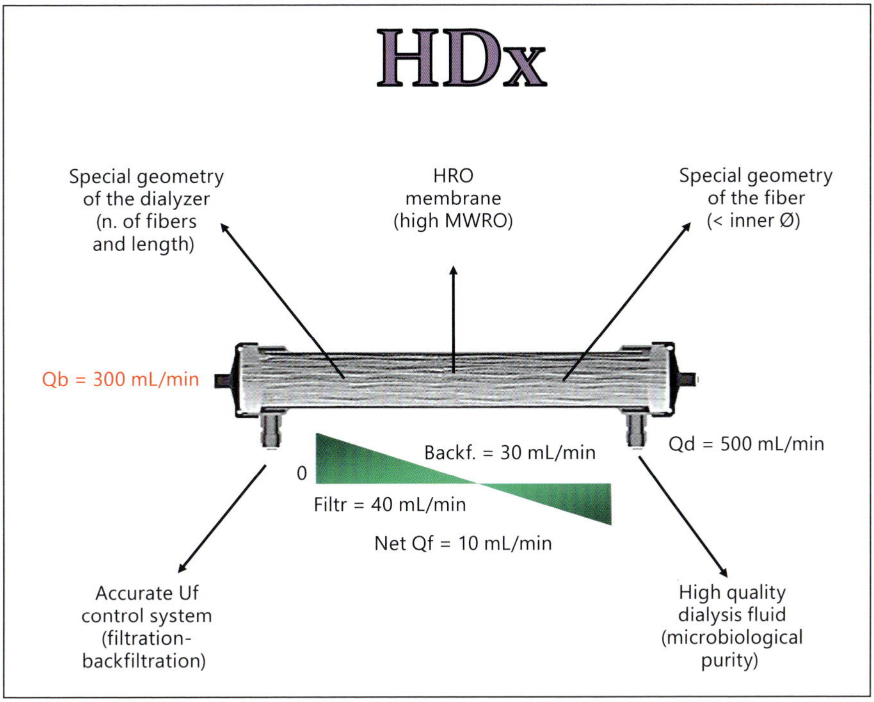

Fig. 5. Requirements to perform HDx and related operational parameters.

There is no need for complex equipment; a blood flow of 300 mL/min or higher and a dialysate flow of 500 mL/min or higher are adequate. Due to the significant amount of backfiltration occurring during treatment, water purity is an important requisite to avoid back transport of contaminants.

For a practical example, we may use the graphs in Figure 1. Recent studies on HDF suggested achieving at least 23 L of ultrafiltration per session [21]. Considering the molecular weight of β-2 M and the sieving properties of a standard HF membrane, total β-2 M convective clearance per session will be 23 × 0.5 = 11.5 L. To achieve the same result with a HRO membrane, where $S = 0.9$, you will only need 12.7 L.

According to our previous studies [22], a 1.3–1.5 m^2 dialyzer equipped with a HRO membrane in a reduced inner diameter configuration should normally generate an average internal filtration of 40 mL/min at zero net filtration. Assuming a dialysis session of 240 min, a *Qb* = 300 mL/min and *Qd* = 500 mL/min, and a scheduled fluid loss from the patient of 1 L/h (16 mL/min), the internal filtration will be 56 mL/min with a backfiltration of 40 mL/min; 56 mL/min will provide a total of 13,440 mL of filtration that, multiplied by 0.9 (*S*), will produce an overall β-2 M clearance of 12.96 L. The result is comparable and even superior to the numbers achievable with HDF, with a simpler treatment and less technical requirements. This should be considered in addition to the possibility of achieving significant clearances of solutes such as λ FLC, whose clearance with HDF is marginal.

Conclusions

The introduction of HRO membranes in the clinical routine has allowed the development of a new concept therapy called HDx. Its simple set-up and application offer the possibility to use it even in patients with suboptimal vascular access or even with an indwelling catheter. The system does not require a particular hardware or unusual nursing skill. The quality of dialysis fluid is, however, mandatory to ensure a safe conduct of the dialysis session [23–25]. This new therapy is likely to modify the outcome of end-stage kidney disease patients, thanks to the enhanced removal of molecules, traditionally retained by current dialysis techniques.

References

1 Vanholder R, de Smet R, Glorieux G, Argilés A, Baurmeister U, Brunet P, Clark W, Cohen G, De Deyn PP, Deppisch R, Descamps-Latscha B, Henle T, Jörres A, Lemke HD, Massy ZA, Passlick-Deetjen J, Rodriguez M, Stegmayr B, Stenvinkel P, Tetta C, Wanner C, Zidek W: Review on uremic toxins: classification, concentration, and interindividual variability. Kidney Int 2003;63:1934–1943.

2 Neirynck N, Vanholder R, Schepers E, Eloot S, Pletinck A, Glorieux G: An update on uremic toxins. Int Urol Nephrol 2013;45:139–150.

3 Miyata T, Jadoul M, Kurokawa K, van Ypersele de Strihou C: Beta-2 microglobulin in renal disease. J Am Soc Nephrol 1998;9:1723–1735.

4 Cianciolo G, Colí L, La Manna G, Donati G, D'Addio F, Comai G, Ricci D, Dormi A, Wratten M, Feliciangeli G, Stefoni S: Is beta2-microglobulin-related amyloidosis of hemodialysis patients a multifactorial disease? A new pathogenetic approach. Int J Artif Organs 2007;30:864–878.

5 Cianciolo G, La Manna G, Colì L, Donati G, D'Addio F, Persici E, Comai G, Wratten M, Dormi A, Mantovani V, Grossi G, Stefoni S: 5-methyltetrahydrofolate administration is associated with prolonged survival and reduced inflammation in ESRD patients. Am J Nephrol 2008;28:941–948.

6 Desjardins L, Liabeuf S, Lenglet A, Lemke HD, Vanholder R, Choukroun G, Massy ZA; European Uremic Toxin (EUTox) Work Group: Association between free light chain levels, and disease progression and mortality in chronic kidney disease. Toxins (Basel) 2013;5:2058–2073.

7 Ronco C, Crepaldi C, Brendolan A, Bragantini L, d'Intini V, Inguaggiato P, Bonello M, Krause B, Deppisch R, Goehl H, Scabardi A: Evolution of synthetic membranes for blood purification: the case of the Polyflux family. Nephrol Dial Transplant 2003;18(suppl 7):10–20.

8 Ronco C: Hemodiafiltration: evolution of a technique towards better dialysis care. Contrib Nephrol 2011;168:19–27.

9 Ronco C: Hemodiafiltration: technical and clinical issues. Blood Purif 2015;40(suppl 1):2–11.

10 Colì L, La Manna G, Comai G, Ursino M, Ricci D, Piccari M, Locatelli F, Di Filippo S, Cristinelli L, Bacchi M, Balducci A, Aucella F, Panichi V, Ferrandello FP, Tarchini R, Lambertini D, Mura C, Marinangeli G, Di Loreto E, Quarello F, Forneris G, Tancredi M, Morosetti M, Palombo G, Di Luca M, Martello M, Emiliani G, Bellazzi R, Stefoni S: Automatic adaptive system dialysis for hemodialysis-associated hypotension and intolerance: a noncontrolled multicenter trial. Am J Kidney Dis 2011;58:93–100.
11 Ward RA: Protein-leaking membranes for hemodialysis: a new class of membranes in search of an application? J Am Soc Nephrol 2005;16:2421–2430.
12 Foley RN, Parfrey PS, Harnett JD, Kent GM, Murray DC, Barre PE: Hypoalbuminemia, cardiac morbidity, and mortality in end-stage renal disease. J Am Soc Nephrol 1996;7:728–736.
13 Kirsch AH, Lyko R, Nilsson LG, Beck W, Amdahl M, Lechner P, Schneider A, Wanner C, Rosenkranz AR, Krieter DH: Performance of hemodialysis with novel medium cut-off dialyzers. Nephrol Dial Transplant 2016;pii:gfw310.
14 Zickler D, Schindler R, Willy K, Martus P, Pawlak M, Storr M, Hulko M, Boehler T, Glomb MA, Liehr K, Henning C, Templin M, Trojanowicz B, Ulrich C, Werner K, Fiedler R, Girndt M: Medium cut-off (MCO) membranes reduce inflammation in chronic dialysis patients – a randomized controlled clinical trial. PLoS One 2017;12:e0169024.
15 Boschetti-de-Fierro A, Voigt M, Storr M, Krause B: Extended characterization of a new class of membranes for blood purification: the high cut-off membranes. Int J Artif Organs 2013;36:455–463.
16 Boschetti-de-Fierro A, Voigt M, Storr M, Krause B: MCO membranes: enhanced selectivity in high-flux class. Sci Rep 2015;5:18448.
17 Ronco C, Brendolan A, Lupi A, Metry G, Levin NW: Effects of a reduced inner diameter of hollow fibers in hemodialyzers. Kidney Int 2000;58:809–817.
18 Fiore GB, Guadagni G, Lupi A, Ricci Z, Ronco C: A new semiempirical mathematical model for prediction of internal filtration in hollow fiber hemodialyzers. Blood Purif 2006; 24:555–568.
19 Fiore GB, Ronco C: Principles and practice of internal hemodiafiltration. Contrib Nephrol 2007;158:177–184.
20 Ronco C: Fluid mechanics and crossfiltration in hollow-fiber hemodialyzers. Contrib Nephrol 2007;158:34–49.
21 Maduell F, Moreso F, Pons M, Ramos R, Mora-Macià J, Carreras J, Soler J, Torres F, Campistol JM, Martinez-Castelao A; ESHOL Study Group: High-efficiency postdilution online hemodiafiltration reduces all-cause mortality in hemodialysis patients. J Am Soc Nephrol 2013;24:487–497.
22 Ronco C, Brendolan A, Feriani M, Milan M, Conz P, Lupi A, Berto P, Bettini M, La Greca G: A new scintigraphic method to characterize ultrafiltration in hollow fiber dialyzers. Kidney Int 1992;41:1383–1393.
23 Weber V, Linsberger I, Rossmanith E, Weber C, Falkenhagen D: Pyrogen transfer across high- and low-flux hemodialysis membranes. Artif Organs 2004;28:210–217.
24 Colì L, Donati G, Cappuccilli ML, Cianciolo G, Comai G, Cuna V, Carretta E, La Manna G, Stefoni S: Role of the hemodialysis vascular access type in inflammation status and monocyte activation. Int J Artif Organs 2011; 34:481–488.
25 Rangel AV, Kim JC, Kaushik M, Garzotto F, Neri M, Cruz DN, Ronco C: Backfiltration: past, present and future. Contrib Nephrol 2011;175:35–45.

Claudio Ronco
Department of Nephrology, Dialysis and Transplantation
International Renal Research Institute of Vicenza (IRRIV), San Bortolo Hospital
Viale Rodolfi 37, IT–36100 Vicenza (Italy)
E-Mail cronco@goldnet.it

La Manna G, Ronco C (eds): Current Perspectives in Kidney Diseases.
Contrib Nephrol. Basel, Karger, 2017, vol 190, pp 134–145 (DOI: 10.1159/000468960)

Mathematical Model of Potassium Profiling in Chronic Dialysis

Mauro Ursino[a] · Gabriele Donati[b]

[a]Department of Electrical, Electronic and Information Engineering, University of Bologna, and [b]Nephrology, Dialysis and Renal Transplant Unit, St. Orsola Hospital, Bologna, Italy

Abstract

Potassium balance is a difficult task in hemodialysis: low potassium in the dialysate is associated with a high risk of sudden cardiac death, whereas excessive dialysate potassium may provoke insufficient removal and hyperkalemia. A better understanding of the problem can be achieved with the use of mathematical models of solute kinetics. This study is aimed at presenting an improved model of solute kinetics and fluid shifts during hemodialysis. It comprises a 2-compartment (intracellular and extracellular) description of sodium, potassium, and urea, including volume fluid shifts induced by osmotic forces. Compared with previous versions, the model also incorporates active $Na^+ – K^+$ transport across the cellular membrane via the sodium-potassium pump. Simulations in chronic conditions, concerning both a 4-h standard hemodialysis and the subsequent 20 h interdialytic period, provide values of solute extracellular concentrations, fluid volumes, and potassium Nernst potential, in agreement with literature. The model predicts that the active $Na^+ – K^+$ transport decreases during the session (mainly due to a decrease in extracellular potassium) but increases in the inter-dialytic period. Extracellular potassium exhibits a rebound in the interdialytic phase, already evident in the first hours. The model also provides preliminary testable predictions on the patterns of intracellular sodium and potassium concentrations, which require further validation based on in vivo data. Particularly, assuming no residual potassium removal from the organism, the intracellular K^+ concentration assumes higher values in the chronic subject compared with the healthy basal conditions. Finally, hemodialysis with profiled potassium (higher in the first half, reduced in the second) is simulated, to point out the advantages compared with the standard session. In perspective, the model can be used to optimize the dialysis treatment on the basis of profiled dialysate concentrations and, with ad hoc in vivo measurements, to provide deeper insight into the mechanisms of internal potassium balance.

Introduction

Achieving a correct potassium balance is a difficult task in hemodialysis. Patients accumulate potassium during the period between dialyses, and hyperkalemia is associated with higher mortality and cardiovascular risks in this population [1]; hence, dialysate potassium concentration should be maintained at a low level to ensure sufficient removal. However, excessive decrease in extracellular potassium is associated with high cardiovascular risks, in particular alterations in cardiac rhythms, sudden cardiac arrest, and sudden cardiac death [2]. There are various evidences that dialysate potassium concentration is often not appropriately adjusted in the modern therapy [3].

Achieving an optimal strategy for potassium removal is a difficult task, which requires a clear understanding of the mechanisms of intracellular and extracellular potassium balance. Actually, the potassium balance is determined by the superimposition of different factors, which are simultaneously affected by treatment. They include potassium exchange across the membrane by passive forces, the active potassium flux into the cell by the $Na^+ - K^+$ pump, the effect of changes in intracellular and extracellular volumes induced by ultrafiltration rates and by osmotic changes, the amount of solutes actually removed by the treatment and those accumulated in the interdialytic phase.

A promising approach to better understand the mechanisms involved and select the optimal dialysate potassium profile consists in the use of kinetic mathematical models. These allow the different elements of the problem to be expressed in rigorous (although simplified) mathematical terms, and the corresponding interactions quantified via theoretical analysis and computer quantitative simulations. Indeed, various such models have been proposed in recent years, with emphasis on different aspects [4, 5].

Among the others, in the past decades we formulated a 2-compartment model of solute kinetics during hemodialysis, which incorporates many aspects delineated above. The model was able to simulate the pattern of plasma solute concentration (Na+, K+, urea) during the four-hour treatment quite well, starting from knowledge of the initial status of the patient (body weight and initial plasma concentrations) [6–9]. The model is actually used to compute the optimal sodium concentration profile in the dialysate during profiled hemodialysis, to counteract the osmotic fall during the early phase of treatment, especially in patients with great incidence of comorbidities [10–12].

The previous model, however, was not adequate for the analysis of intracellular potassium. In particular, since the emphasis was mainly on extracellular solutes, the potassium exchange at the cellular membrane was over-simplified, assuming a constant ratio between intracellular and extracellular concentrations at equilibrium. However, as pointed out by Sterns et al. [13], during potassium loading, the changes in intracellular potassium concentration are proportionally less than those in extracellular potassium concentration. Shifts in the intracellular versus extracellular fluid volumes and the modifications in the action of the active pump may explain this phenomenon.

The objective of this study is to present a significantly improved version of the previous model, which accounts for the effect of the sodium-potassium pump. A similar bi-compartment model of potassium kinetics, including the sodium-potassium pump, has been proposed by Ciandrini et al. [4] recently to optimize dialysate potassium profiling. However, this model accounts only for the dependence of the pump on extracellular potassium and, more importantly, does not incorporate the interaction between potassium and other solutes in the intracellular and extracellular fluids. Conversely, the model presented here introduces more accurate equations for the sodium-potassium pump, and accounts for the interactions between potassium kinetics and the kinetics of other solutes (especially sodium), which affect both the cell volume and the action of the pump.

The main aspects of the model are presented in a qualitative way, laying emphasis on new equations (concerning the sodium-potassium pump). Model results, concerning both the intradialytic and interdialytic phases, are presented, and compared with some data in the literature. Finally, strategic lines for the use of this model in profiling potassium are delineated and discussed.

Method

Since most equations have been already presented in previous studies [6–9], only the qualitative aspects will be described. New equations (concerning the sodium-potassium exchange at the membrane) are given in detail.

The model incorporates a 2-compartment description of body fluids and of solute kinetics (interstitial fluid and intracellular fluid).

Body Fluid Model

During hemodialysis, the intracellular fluid volume may change as a consequence of water flow through the cellular membrane. This flow depends on the difference between the osmotic activities in the intracellular and the interstitial fluid. In the model, osmotic activities are calculated from the sum of the main solute concentrations, by means of the kinetics equations including additional masses, not explicitly taken into consideration (see [6–9] for more detail).

The volume change of the interstitial fluid depends not only on the water flow through the cellular membrane, but also on the ultrafiltration rate at the dialyzer. The latter is an input quantity for the model.

Solute Kinetics

In the version used here, the model incorporates the kinetics of the following main solutes: sodium, potassium, urea, while the overall mass of the remaining solutes (bicarbonate, H^+, CO_2, chloride, and proteins) is assumed to remain constant (since some of them are eliminated while others are provided during hemodialysis). These solutes can be easily included in future works, by additional kinetic equations.

The mass of each solute is assumed to be distributed in 2 pools, that is, the intracellular and extracellular. The latter pool is equal to the sum of interstitial fluid and plasma. Interstitial fluid and plasma concentrations, however, are a little different, as a consequence of the Gibbs–Donnan equilibrium, and of differences between plasma and plasma water [14].

Equations are based on the conservation of mass for each solute at the cellular and at the dialyzer membranes.

The changes of solute masses in the intracellular fluid per unit time depend on three terms: solute mass flow through the cellular membrane via passive mechanisms (assumed proportional to the difference between intracellular and extracellular concentrations), a constant production rate, and the rate at which the solute is transported by the active pumps.

The change of solute mass in the extracellular fluid depends, besides mass flow at the cellular membrane, also on solutes exchanged at the dialyzer. The latter includes both convective and diffusive transports [7, 8].

Equations for Sodium and Potassium

In the following, new equations, concerning active transport by the sodium-potassium pump, are presented. The meaning of all symbols is presented in Table 1, together with basal parameter values.

Intracellular potassium kinetics:

$$\frac{dM_{K,ic}(t)}{dt} = k_K\left(c_{K,ex}(t) - c_{K,ic}(t)\right) + 2J_p(t) + G_K . \tag{1}$$

Extracellular potassium kinetics:

$$\frac{dM_{K,ex}(t)}{dt} = -k_K\left(c_{K,ex}(t) - c_{K,ic}(t)\right) - 2J_p(t) - J_K(t). \tag{2}$$

The expression for potassium exchanged at the dialyzer, $J_K(t)$, is

Table 1. Meaning of symbols and basal values

Symbol	Significance	Basal value
$c_{K,\,ic}$	Intracellular potassium concentration	140 mEq/L
$c_{K,\,ex}$	Extracellualar potassium concentration	4 mEq/L
G_K	Potassium production rate	65 mEq/day
$c_{K,\,D}$	Potassium concentration in the dialysate	2 mEq/L
$c_{Na,\,ic}$	Intracellular sodium concentration	14 mEq/L
$c_{Na,\,ex}$	Extracellualar sodium concentration	142 mEq/L
$c_{Na,\,D}$	Sodium production rate	154 mEq/day
G_{Na}	Sodium concentration in the dialysate	140 mEq/L
J_p	Cycle-flux of the $Na^+ - K^+$ pump	1.3 mEq/min
Q_F	Ultrafiltration rate	8.33 mL/min (0.5 L/h)
Q_B	Blood flow rate at the dialyzer	0.3 L/min
D	Dialysance	0.241 L/min
E_K	Potassium Nernst potential	–94.9 mV

$$J_k(t) = Q_F(t)\,c_{K,ex}(t) + D\left(1 - \frac{Q_F(t)}{Q_B}\right)\left(c_{K,ex}(t) - c_{K,D}(t)\right). \tag{3}$$

Similar equations hold for sodium.

Intracellular sodium kinetics:

$$\frac{dM_{Na,ic}(t)}{dt} = k_{Na}\left(c_{Na,ex}(t) - c_{Na,ic}(t)\right) - 3J_p(t) + G_{Na}\,. \tag{4}$$

Extracellular sodium kinetics:

$$\frac{dM_{Na,ex}(t)}{dt} = -\,k_{Na}\left(c_{Na,ex}(t) - c_{Na,ic}(t)\right) + 3J_p(t) - J_{Na}(t). \tag{5}$$

Sodium exchanged at the dialyzer, $J_{Na}(t)$:

$$J_{Na}(t) = Q_F(t)\,c_{Na,ex}(t) + D\left(1 - \frac{Q_F(t)}{Q_B}\right)\left(c_{Na,ex}(t) - c_{Na,D}(t)\right). \tag{6}$$

Finally, the term J_p in the previous equations is the so-called "cycle-flux," computed by the state diagram of process of transport in the sodium-potassium pump mediated by $Na^+ - K^+$-ATPase [15]. Although the final expression is quite complex, the following main dependence on concentrations can be summarized:

$$J_p(t) = k_{p1}\, c^3_{Na,\,ic}(t)\, c^2_{K,\,ex}(t) - k_{p2}\, c^3_{Na,\,ex}(t)\, c^2_{K,\,ic}(t), \tag{7}$$

where k_{p1} and k_{p2} are complex functions of the rate constants and ligand concentrations. However, for the sake of simplicity, in the present model these are considered as 2 constant parameters, that are assigned on the basis of physiological considerations.

Equations similar to (1)–(3) hold for the urea too, but without the term representing pump transport.

Finally, we also computed the K^+ Nernst potential as follows:

$$E_K = -\frac{RT}{F}\ln\left(\frac{c_{K,ic}(t)}{c_{K,ex}(t)}\right) \tag{8}$$

with $RT/F = 26.7$ mV at 37°C.

Assignment of Model Parameters

All parameters of the model have been given in agreement with the literature or our previous work [6–9]. However, four parameters are new, depending on the sodium-pump equation. They are k_K, k_{Na}, k_{p1}, and k_{p2}.

In order to provide a value for these parameters, we first looked for a value of the "cycle-flux" J_p in normal conditions (i.e., when all concentrations have their basal values typical of a healthy subject). It is worth noting that potassium mass transport by the pump is $2J_p$ (equation 2), while sodium mass transport by the pump is $3J_p$ (equation 5).

Different values for the normal potassium transport by the pump are given in the literature, which may differ greatly. Here, we used a value taken from the nonlinear model by Ciandrini et al. [4]. The advantage is that, with their model, the authors provided a fitting to experimental K^+ plasma concentrations during dialysis and post-dialysis rebound, which indirectly support this choice. We approximately have, for the active potassium transport: $2J_p = 2.6$ mmol/min. From this value, we computed the exchange parameters, k_K, k_{Na}, assuming that equations (1)–(5) are at equilibrium in a healthy subject (i.e., assuming that solute production rates are balanced by renal function). This can be simulated by assuming $G_K = 0$ and $G_{Na} = 0$ in equations (1) and (4), and $J_K = 0$ and $J_{Na} = 0$ in equations (2) and (5). However, equations are not in equilibrium during the intradialytic phase for a patient with total renal failure, since $G \neq 0$, and so concentrations increase in the inter-dialytic period. Finally, values for the parameters k_{p1} and k_{p2} were given assuming that the first term in equation (7) in normal conditions is twice the second. This is quite an arbitrary choice. However, we found that the final results are only scarcely affected by a change in the ratio k_{p1}/k_{p2}.

Results

First, we simulated a repeated series of hemodialysis sessions, each with a 4-h duration, repeated every 2 days, starting from a normal condition (i.e., an healthy equilibrium state). For the sake of simplicity, we did not consider the difference between the short and long interdialytic periods. This may be easily included in future works. After several consecutive dialysis sessions (approximately 10), the model settles at a steady state condition, characterized by the same values of all quantities at the beginning of each session (i.e., the initial transient period, characterized by a change in the solute masses, is exhausted).

The pattern of the main quantities of the model (intracellular and extracellular sodium, potassium and urea concentrations, intracellular and extracellular volumes, pump cycle flux, and potassium Nernst potential) during one intradialytic and the subsequent interdialytic phase in this steady state conditions are reported in Figure 1.

Results show that the intracellular potassium concentration exhibits just minor changes during the intradialytic phase and in the subsequent post-dialysis period. However, at the hemodialysis regimen, the intracellular potassium concentration settles at about 150 mEq/L, compared with the normal basal value in the model that was as low as 140 mEq/L. This signifies that a repeated treatment with a potassium concentration in the dialysate equal to 2 mEq/L leads to a condition of intracellular hyperkalemia (i.e., following 20 days, with 10 treatments, intracellular potassium rose from 140 to 150 mEq/L to settle at this quite constant level). The extracellular potassium shows large percentage changes during the session and in the interdialytic period: these agree with the data reported in the literature [16, 17]. As a consequence of the extracellular potassium changes, the K^+ Nernst potential decreases dramatically during the session from approximately –90 mV to approximately –100 mV. The "flux rate" of the sodium-potassium pump decreases during the session and increases during the interdialytic period. This pattern mainly reflects the large percentage changes in extracellular potassium concentration.

The patterns of extracellular sodium concentration and extracellular urea agree with those commonly observed during hemodialysis, showing a monotonic decrease during the session, and a subsequent slow recovery during the interdialytic phase.

Extracellular volume decreases during the session, while intracellular volume increases during the first 2 h of treatment, reflecting a fluid shift due to osmotic forces. Both volumes increase in the interdialytic phase due to fluid intake.

Subsequently, we simulated hemodialysis sessions with a profiled potassium in the dialysate, higher in the first half (c_{Kd} = 3.5 mEq/L) and lower in the second half (c_{Kd} = 1.5 mEq/L). This kind of "profiled potassium" ensures almost the same potassium removal as the previous simulations, but maintains a higher K+ extracellular concentration and Nernst potential during the first hours of treatment. A summary of the latter results, during one hemodialysis session and the subsequent 4 h of recovery, is shown in Figure 2. The time patterns of fluid volumes and urea concentrations are not shown, since they exhibit only minor changes compared to those in Figure 1.

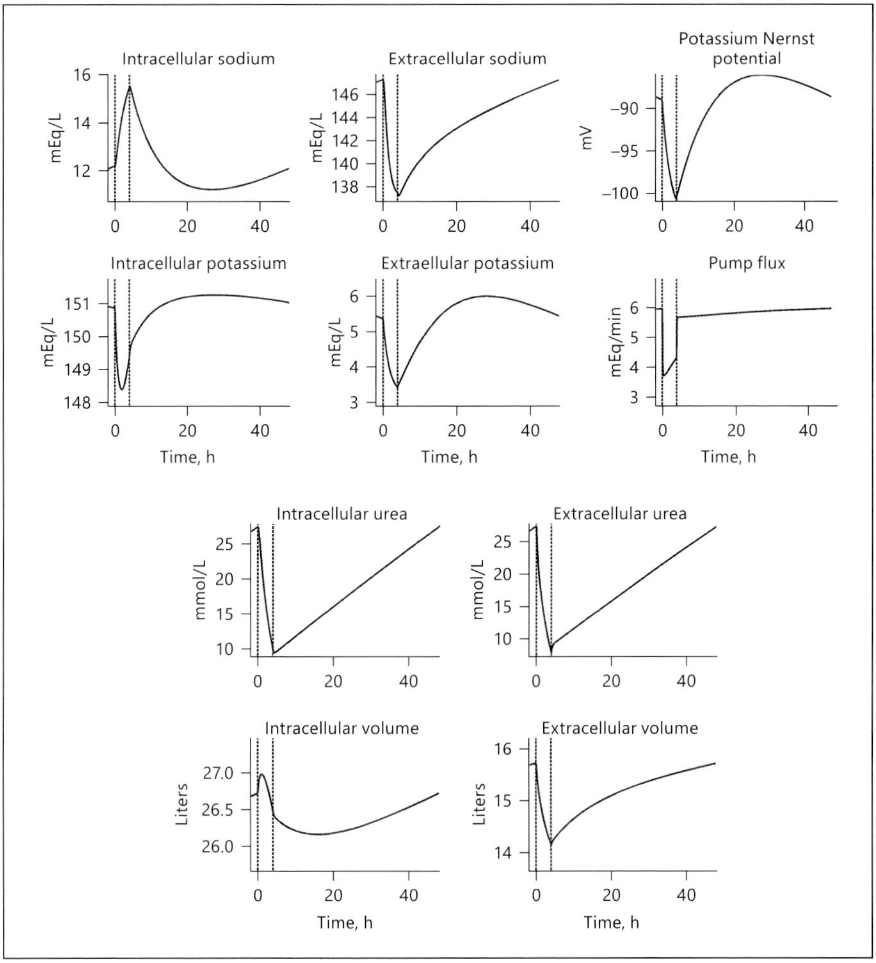

Fig. 1. Time pattern of the main quantities in the model: intracellular and extracellular potassium, sodium, and urea concentrations, "cycle flux" of the sodium potassium pump (equal to half the active flux of potassium and one third of the active flux of sodium), Nernst potential of potassium across the cellular membrane, intracellular and extracellular volumes. The period between the two vertical dashed lines refers to a standard 4-h dialysis (dialysate composition: sodium 140 mEq/L; potassium 2 mEq/L; ultrafiltration rate: 500 mL/h). The subsequent period is the interdialytic phase. The model is in steady-state conditions simulating a chronic patient.

Discussion

The model described in this work represents a first attempt to analyze potassium balance in hemodialysis, taking into account the different phenomena that can be involved. The focus is not only on the extracellular/plasmatic concentration,

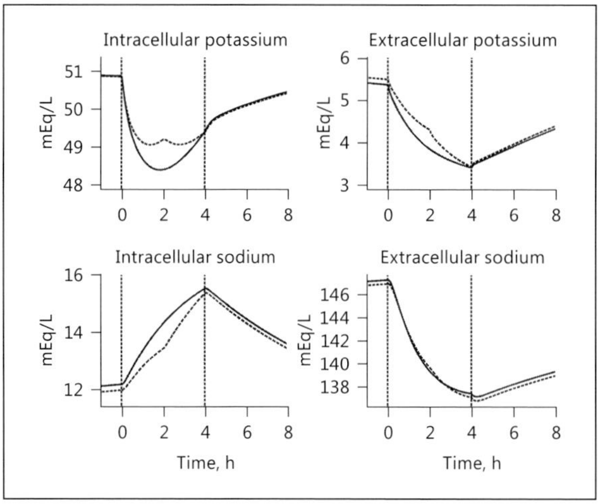

Fig. 2. Time pattern of intracellular and extracellular sodium, and intracellular and extracellular potassium concentrations computed with the model assuming first a standard hemodialysis session (the same values as in Figure 1: continuous line) and then a potassium profiled session (potassium concentration in the dialysate as high as 3.5 mEq/L in the first 2 h and as low as 1.5 mEq/L in the last 2 h: dashed line). The period between the 2 vertical lines refers to a 4-h dialysis. Four hours in the post-dialysis period are also shown.

but also on the intracellular pool, where about 98% of the overall potassium mass in located. The model represents an extension of previous models developed in past years [6–9]; however, those models were especially focused on the optimization of sodium balance and extracellular volume, in an effort to avoid typical intradialytic symptoms (like cardiovascular hypotension and disequilibrium syndrome); hence, less attention was given to potassium. Although an important model including the sodium-potassium pump was presented recently [4], the present study incorporates many different aspects that were not included in that model, such as the sodium and urea effects on osmolarity, the volume fluid shift, a more complete equation for the pump.

The present model includes many parameters that were given empirically, on the basis of physiological knowledge and previous modeling studies, in an effort to simulate the behavior of a prototypical subject. Of course, future work will be focused on more accurate parameter estimation, based on ad hoc measurements in individual patients. Nevertheless, the present results agree with several results in the literature quite well. In particular, the pattern of plasmatic potassium concentration (which is computed as

$$c_{K,pl}(t) = \frac{F_p}{\alpha_K} c_{K,ex}(t),$$

where F_p is the plasma water fraction and α_K is the Gibbs–Donnan ratio, equal to 0.94 and 0.95, respectively) agrees quite well with the patterns reported by some authors in the intradialytic period [4, 5, 16, 18]. Moreover, our simulations show a significant rebound of plasmatic potassium concentration in the inter-dialytic phase, which is quite similar to that described by Blumberg et al. [16]. These authors reported that there is a rapid post-dialysis rebound of plasma K. They also observed a strong correlation between the pre-dialysis and 6-h post-dialysis plasma concentrations, according to the equation $K_6 = 0.658\ K_0 + 1.294$ (where K_0 is pre-dialysis and K_6 is 6-h post-dialysis concentration). With this equation, using our potassium concentration at the beginning of dialysis (K_0 = 5.30 mEq/L), we obtain a theoretical value at 6 h, K_6 = 4.79 mEq/L, which is quite close to the value actually provided by model simulation ($c_{K,\,pl}$ (6 h) = 4.65 mEq/L). A slight difference between our results and those by Blumberg et al. [16] concerns the value at 16 h after the treatment: our model suggests a moderate overshoot compared with the pre-dialysis plasma potassium ($c_{K,\,pl}$ (16 h) = 5.74 mEq/L), whereas Blumberg et al. [16] report that potassium is still moderately reduced.

Other model predictions are in good agreement with the literature. In particular, the overall K removal in our model is about 113 mEq, which agrees with Blumberg et al. [16] and is a little higher than in Ciandrini et al. [4]. More important, several authors state that a large portion of K removal arises from the intracellular compartment (58% according to Blumberg et al. [16], more than 50% looking at Fig. 6 in Ciandrini et al. [4]). We computed the K mass reduction during our standard simulation, and found that about 68% of potassium is removed from the intracellular compartment and just 32% from the extracellular.

Another interesting prediction of the model is that the percentage changes in intracellular K concentration, both during the session and in the subsequent interdialytic phase, are quite small compared with the percentage extracellular potassium concentration changes. Previous studies that measured potassium concentration within red blood cells reported a small (4%) [19] or no reduction at the end of treatment [17].

A critical point of our simulations is that, to reach a stable steady state potassium removal, equal to the potassium production rate (about 65 mEq/day in our simulations), potassium concentration in the intracellular pool must settle at a value considerably higher than normal (we started with an intracellular "normal" potassium as low as 140 mEq/L; the steady state hemodialysis condition suggests a value around 150 mEq/L). The reason is that the force that governs potassium removal is the K^+ gradient between the intracellular and extracellular fluids, and this must be strong enough to counterbalance the pump rate, which is increased during the inter-dialysis period (especially due to the increase in intracellular sodium). Possibly, a smaller value for potassium concentration in the intracellular pool can be achieved, in future model applications, by assuming a residual potassium removal rate in the chronic patient (for instance, through the intestinal tract) [20]. The latter aspect has not been incorporated in the present model and requires further validation tests.

Results obtained with different patterns of potassium concentration in the dialysate suggest that an improved pattern of extracellular potassium concentration can be obtained by profiling potassium, with a higher concentration in the first half of the session, and a reduced value at the end. In this regard, our simulations substantially agree with those by Ciandrini et al. [4] and open interesting potential model applications in future work.

The present model exhibits some limitations, which will be the target of future work. First, some parameters require ad hoc assignment on the basis of real patients' data. To this end, we think that intracellular potassium concentration must be measured both in the intra- and interdialytic phases in a number of exemplary sessions, to better validate the model. Other solutes (not incorporated here, such as chloride, bicarbonate, creatinine) can affect osmotic changes and volume shifts between the intracellular compartments. In the present study, we assumed that the overall mass of these solutes is constant. Future model versions will incorporate the kinetics of them. K^+ kinetics can be significantly affected by pH changes [13], a problem that deserves future attention. Finally, the model can be linked with cardiovascular regulation models (as in [21] and [22]) to better understand the K^+ effect on heart.

References

1 Charytan DM, Foley R, McCullough PA, Rogers JD, Zimetbaum P, Herzog CA, Tumlin JA; MiD Investigators and Committees: Arrhythmia and sudden death in hemodialysis patients: protocol and baseline characteristics of the monitoring in dialysis study. Clin J Am Soc Nephrol 2016;11:721–734.

2 Kovesdy CP, Regidor DL, Mehrotra R, Jing J, McAllister CJ, Greenland S, Kopple JD, Kalantar-Zadeh K: Serum and dialysate potassium concentrations and survival in hemodialysis patients. Clin J Am Soc Nephrol 2007; 2:999–1007.

3 Pun PH, Lehrich RW, Honeycutt EF, Herzog CA, Middleton JP: Modifiable risk factors associated with sudden death within hemodialysis clinics. Kidney Int 2011;79:218–227.

4 Ciandrini A, Severi S, Cavalcanti S, Fontanazzi F, Grandi F, Buemi M, Mura C, Bajardi P, Badiali F, Santoro A: Model-based analysis of potassium removal during hemodialysis. Artif Organs 2009;33:835–843.

5 Agar BU, Culleton BF, Fluck R, Leypoldt JK: Potassium kinetics during hemodialysis. Hemodial Int 2015;19:23–32.

6 Ursino M, Colì L, Dalmastri V, Volpe F, La Manna G, Avanzolini G, Stefoni S, Bonomini V: An algorithm for the rational choice of sodium profile during hemodialysis. Int J Artif Organs 1997;20:659–672.

7 Ursino M, Colì L, Brighenti C, De Pascalis A, Chiari L, Dalmastri V, La Manna G, Mosconi G, Avanzolini G, Stefoni S: Mathematical modeling of solute kinetics and body fluid changes during profiled hemodialysis. Int J Artif Organs 1999;22:94–107.

8 Ursino M, Colì L, Brighenti C, Chiari L, De Pascalis A, Avanzolini G: Prediction of solute kinetics, acid-base status, and blood volume changes during profiled hemodialysis. Ann Biomed Eng 2000;28:204–216.
9 Ursino M, Colí L, Magosso E, Capriotti P, Fiorenzi A, Baroni P, Stefoni S: A mathematical model for the prediction of solute kinetics, osmolarity and fluid volume changes during hemodiafiltration with on-line regeneration of ultrafiltrate (HFR). Int J Artif Organs 2006;29:1031–1041.
10 Colì L, Ursino M, Donati G, Cianciolo G, Soverini ML, Baraldi O, La Manna G, Feliciangeli G, Scolari MP, Stefoni S: Clinical application of sodium profiling in the treatment of intradialytic hypotension. Int J Artif Organs 2003;26:715–722.
11 Colì L, La Manna G, Comai G, Ursino M, Ricci D, Piccari M, Locatelli F, Di Filippo S, Cristinelli L, Bacchi M, Balducci A, Aucella F, Panichi V, Ferrandello FP, Tarchini R, Lambertini D, Mura C, Marinangeli G, Di Loreto E, Quarello F, Forneris G, Tancredi M, Morosetti M, Palombo G, Di Luca M, Martello M, Emiliani G, Bellazzi R, Stefoni S: Automatic adaptive system dialysis for hemodialysis-associated hypotension and intolerance: a noncontrolled multicenter trial. Am J Kidney Dis 2011;58:93–100.
12 Cianciolo G, Colí L, La Manna G, Donati G, D'Addio F, Comai G, Ricci D, Dormi A, Wratten M, Feliciangeli G, Stefoni S: Is beta2-microglobulin-related amyloidosis of hemodialysis patients a multifactorial disease? A new pathogenetic approach. Int J Artif Organs 2007;30:864–878.
13 Sterns RH, Cox M, Feig PU, Singer I: Internal potassium balance and the control of the plasma potassium concentration. Medicine (Baltimore) 1981;60:339–354.
14 Hall JE: Compartimenti liquidi dell'organismo: liquidi extracellulare e intracellulare, edema; in Guyton AC, Hall JE (eds): Fisiologia Medica 12° Edizione. Milano, Elsevier, 2012, pp 281–296.
15 Hernández JA, Cristina E: Modeling cell volume regulation in nonexcitable cells: the roles of the Na+ pump and of cotransport systems. Am J Physiol 1998;275:C1067–C1080.
16 Blumberg A, Roser HW, Zehnder C, Müller-Brand J: Plasma potassium in patients with terminal renal failure during and after haemodialysis; relationship with dialytic potassium removal and total body potassium. Nephrol Dial Transplant 1997;12:1629–1634.
17 Buemi M, Aloisi E, Coppolino G, Loddo S, Crascì E, Aloisi C, Barillà A, Cosentini V, Nostro L, Caccamo C, Floccari F, Romeo A, Frisina N, Teti D: The effect of two different protocols of potassium haemodiafiltration on QT dispersion. Nephrol Dial Transplant 2005;20:1148–1154.
18 Basile C, Libutti P, Lisi P, Teutonico A, Vernaglione L, Casucci F, Lomonte C: Ranking of factors determining potassium mass balance in bicarbonate haemodialysis. Nephrol Dial Transplant 2015;30:505–513.
19 Rombolà G, Colussi G, De Ferrari ME, Frontini A, Minetti L: Cardiac arrhythmias and electrolyte changes during haemodialysis. Nephrol Dial Transplant 1992;7:318–322.
20 Allon M: Disorders of potassium metabolism; in Greenberg A, Cheung AK, Coffman TM, Falk RJ, Jennette JC (eds): Primer on Kidney Diseases, ed 3. San Diego, Academic Press, 2001, pp 98–107.
21 Ursino M, Innocenti M: Modeling arterial hypotension during hemodialysis. Artif Organs 1997;21:873–890.
22 Ursino M, Innocenti M: Mathematical investigation of some physiological factors involved in hemodialysis hypotension. Artif Organs 1997;21:891–902.

Mauro Ursino
Department of Electrical, Electronic and Information Engineering
University of Bologna
Viale Risorgimento 2, IT–40136 Bologna (Italy)
E-Mail mauro.ursino@unibo.it

La Manna G, Ronco C (eds): Current Perspectives in Kidney Diseases.
Contrib Nephrol. Basel, Karger, 2017, vol 190, pp 146–155 (DOI: 10.1159/000468961)

Home Hemodialysis: What Is Old Is New Again

Peter G. Kerr[a] • Juli Jaw[a, b]

Department of Nephrology, [a]Monash Health and [b]Monash University, Clayton, VIC, Australia

Abstract

Home hemodialysis (HD) is undergoing a resurgence. A major driver of this is economics, however, providers are also encouraged by a combination of excellent patient outcomes and patient experiences as well as the development of newer technologies that offer ease of use. Home HD offers significant advantages in flexible scheduling and the practical implementation of extended hours dialysis. This paper explores the reasons why home HD is making a comeback and strives to offer approaches to improve the uptake of this dialysis modality.

Brief History

In the early era of dialysis, home hemodialysis (HD) was a common occurrence – especially in the UK, parts of the US, Australia and New Zealand. With the spread of dialysis facilities and the increasing complexity of HD machines and related technology, home HD declined around the world. By the early 2000s, only Australia, New Zealand, and Turkey had significant proportions of home HD (about 11.1% of all dialysis in Australia, 13.7% in New Zealand, and 11.1% in Turkey) [1].

However, the last 15 or so years have seen a return to favor of home HD in some parts of the world. This has tended to occur in pockets, for example – parts of the UK, small pockets in Europe and Scandinavia, and some regions of the USA and Canada. Outcomes for home HD have always been excellent and home HD is cheaper for most economies than center-based therapies. This study explores the pros and cons of home HD from the perspective of an unabashed enthusiast.

Why It Faded?

In the initial era of HD, many patients utilized home HD because the alternative was a long commute to a dialysis facility. In some countries, such as Australia this may have meant many hundreds of kilometers of travel per week. Gradually, over time, more and more dialysis facilities were established, many of these remote from hospitals but closer to the patients, making it more attractive to commute to the facilities where experienced staff took the responsibility for the dialysis procedure.

In some countries, money was to be made from in-patient dialysis – both for facilities and doctors. This led to incentives for facility-based therapies and significant disincentives for home-based therapies, whether they be HD or peritoneal dialysis (PD) [2]. Compounding this, funding models for dialysis were not necessarily supportive of home therapies. This trend had an unintended consequence: the teaching and support skills for home HD also waned and were lost to many services. From an equipment point of view, the concept of one machine, one patient sounds expensive although in most developed countries staff costs far outweigh the cost of machines.

In some countries, there was also a growing fear of legal liabilities for a complex treatment being performed without the direct supervision of a clinician. This, in association with patient concerns about the complexity of the treatment and their role in self-management and especially self-needling provided some disincentives, particularly if not fully supported and encouraged by the clinical team. Training for home HD takes patience and investment of time by the home HD nursing staff (and by the patient and carer) – the lack of available trained staff and the lack of support for training time became a problem. In addition, many jurisdictions would not allow home HD without the presence of a caretaker/partner to attend to potential dialysis incidents and emergencies.

Another factor has been the increase in transplantation. With the change in acceptance of living donors from strictly related family members to non-related family and friends, ABO-incompatible donors and the evolution of paired kidney exchange programs, there has been a significant shift in living donation. In addition, schemes to improve cadaveric donation have been instituted in many countries – these include "the Spanish model" [3] and opt-out schemes [4]. Overall, renal transplantation rates have almost doubled in many countries – in 2001, the rates varied from 10 to 55 pmp and in 2015, the rates were more consistently around 50 pmp across a broader range of countries [5]. Given that a large proportion of home HD patients are ideal transplant recipients, this has had a significant impact on home dialysis programs.

Now It Is Back

Since the early 2000s, home HD has had a degree of resurgence. Even in stronghold countries such as Australia and New Zealand, the rate of home HD has risen over the ensuing years from a nadir in 2005 [6]. The UK has also shown an increase. While the largest dialysis market – the USA – still has a low proportion of home HD, the absolute number has reached more than 5,000 home HD patients in 2015 [5].

What Has Changed?

Much of the drive to increase home HD has been economic. The quickly expanding pool of dialysis patients in many countries has stretched available resources – including available physical facilities, staff, and hospital budgets. The ongoing need to build more and more dialysis facilities has been beyond the reach even for many developed countries. Whilst home HD is typically "one machine, one patient," machines are significantly cheaper than staff in most developed countries. A single dialysis machine used in the home may be functional for 10 years, minimizing the impact of this cost [7]. This, combined with reports of excellent outcomes, has prompted administrators to ask the Nephrology community why they are not pursuing this modality more. Indeed, in countries such as Australia where a national health funding scheme covers the cost of dialysis, definitive financial incentives are now offered to promote home dialysis (e.g., in the State of Victoria, hospitals receive an incentive payment of AUD $10,000 for every home HD patient). Patients in some regions also receive financial assistance via the Department of Health to help defray the additional electricity and water charges [8].

Changes to technology have also helped. Individual water treatment plants have been streamlined and operate more efficiently and quietly making them more acceptable in the home. Dialysis machines specifically designed for home use have evolved, with simpler patient control panels, and simpler set-up and circuitry (e.g., cartridge-based lines). The introduction of the NxStage machine certainly had a major impact in the US, with this machine dominating the home HD market in that country [9]. Much of the expansion of home HD in recent years in the US can be traced to the introduction of the NxStage machine. This machine has a simple set-up and quick on-off procedure and is relatively quick to learn to operate; it also has a degree of portability. These factors together made it very attractive in a home HD naïve environment.

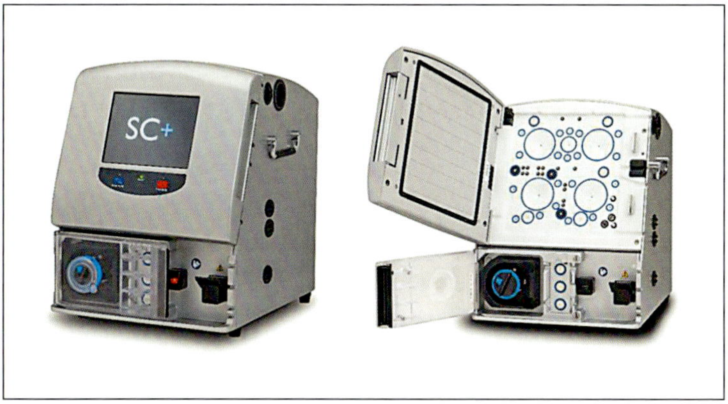

Fig. 1. An example of the new machines becoming available for home HD, with a small footprint, overall reduced size, and cassette-type fluid pathway. Permission for publication granted – Quanta Fluid Solutions.

In more recent years, a number of new home-friendly dialysis machines have been developed, although not all are yet available. These include the Quanta, Physidia, and C-PAK machines – all of which incorporate cassette technology, quick set-up, and portability (Fig. 1). All are single-pass, normal dialysate flow rate machines (unlike NxStage which utilizes low dialysate flow rates). It will be interesting to see their impact over the next few years.

Another development in recent years has been the fall-out from PD. As a measure to increase home dialysis in general, PD has also experienced a resurgence, with PD-first approaches adopted in some countries [10]. However, the technique survival of PD is commonly less than for HD modalities [10]. Whilst this might have been expected to result in a number of patients who were already accustomed to dialyzing at home to have transitioned to HD – in most jurisdictions, this trend has not been large, with the majority of failed PD patients actually drifting towards facility-based HD. This may be because they were unwell at the time of transition. Attempting to predict impending PD technique failure ahead of time and transitioning the patients whilst they are still well may help to address this.

A potential negative impact from new technology relates to monitoring of home HD patients. With WiFi and Bluetooth technology have come various approaches to the monitoring of patients' progress at home. This potentially includes simple parameters such as BP and weight but may also include machine parameters (blood flow rate, achieved ultrafiltration etc.), blood leak detection, and even direct visualization of the session. Whilst it is easy to claim these improve the safety of the sessions, they also increase the intrusiveness of the "hospital" into the lives of the patients. Many patients opted for home HD to avoid the excess medical intervention in their lives. Finding a balance between safety and autonomy will likely become increasingly more important with the rapid changes in available technology [11].

Outcomes

On paper, home HD is usually associated with excellent mortality outcomes. In Australia/New Zealand, the mortality of home HD patients is reported at 4.8/100 patient years, compared to 15.0/100 patient years for the general (non-home) HD population and 15.7/100 patient years for PD (derived from 6 and personal communication, S McDonald, ANZDATA). However this data does require careful interpretation, as home HD patients tend to have several favorable features: they tend to be younger, are more likely to be working, have less diabetes and other comorbidities, and have greater engagement with their own care. Self-care, in dialysis and other settings, is associated with better outcomes [12].

One analysis that has attempted to control many of these factors is that of Marshall et al. [13], using the ANZDATA registry from Australia and New Zealand. This has the advantage of being an all-inclusive registry that has a relatively high percentage of home HD patients. Using marginal structural models, this analysis reinforces the advantages gained by home HD, especially using intensive dialysis regimens which had a hazard ratio for death of 0.56 compared to conventional facility-based HD.

A major factor in improved outcomes for home HD, at least in some countries, is the utilization of greater hours per week dialysis – whether it be short daily, or extended hours (± nocturnal) dialysis. This results in improved biochemical control and especially volume control. Indeed, in the Marshall analysis of ANZDATA, it was the extended hours home HD patients that demonstrated the greatest mortality benefit [13]. Buchanan et al. [14] have proposed that the major determinant of intra-dialytic cardiac stunning is volume control and resultant ultrafiltration rate. Thus, the extended hours pattern of home HD allows for slower UF rates resulting in the lowest rate of cardiac stunning [15].

Apart from data related to more frequent dialysis, there is little data on fistula outcomes for home HD patients. Most home HD patients needle themselves – and interestingly, once comfortable with this, far prefer to needle themselves. It remains uncertain whether this may potentially result in worse (or better) fistula outcomes. In one Australian study (using predominantly extended hours, alternate day regimens), access event-free rates in 286 patients were 68% at 2 years and 61% at 5 years [16]. More data are required to determine if this needs particular attention.

Advantages

There are several significant advantages to home HD apart from the obvious economic advantage. As mentioned above, the mortality outcomes are excellent, although it must be interpreted carefully. Beyond this, a significant advantage is treatment flexibility. Even utilizing a regular, 3 times a week schedule, patients are not locked into a facility's schedule and can manipulate work and social activities according to their own requirements.

A major scheduling advantage is the capacity to avoid long breaks (e.g., on a Mon/Wed/Fri schedule). In the author's own unit, most home HD patients dialyze on a strictly alternate day schedule, thus avoiding the excess solute and volume accumulation associated with the "weekend" long breaks. This 16% increase in episodes of care is affordable, especially given the reduced costs of home HD, and provides significant advantages in volume control, in particular and may avoid the high mortality seen after long breaks [16, 17].

Nocturnal schedules are predominantly confined to home dialysis. Although possible in facilities, the costs in a staffed facility can become prohibitive. Further, facility-based nocturnal programs tend to maintain a 3-times a week schedule, thus losing some of the advantages of avoiding long breaks. On the contrary, nocturnal HD at home is the norm for many units. In our own unit, 90% of home HD patients undergo nocturnal HD (on an alternate night schedule). Nocturnal HD in this setting incurs no additional costs compared to regular day-time HD, yet offers the advantages of extended hours dialysis with slow ultrafiltration rates and optimal solute control. Whilst the few RCTs in this area have failed to demonstrate a clear advantage in (surrogate) outcomes [18], Australian and North American registry data and personal experience are overwhelmingly in favor of these schedules [13, 19].

Of course, the alternative feasible approach to home HD is to undertake short-daily schedules. This is not widely practiced in Australia/New Zealand but is popular, in conjunction with the NxStage technology, in the US. Outcomes are variable – in the FHN trials, short daily dialysis was associated with improvement in the composite end-point of death and left-ventricular hypertrophy [20] (admittedly as predominantly in-centre dialysis), but registry-type data does not show as much of a clear advantage as extended hours dialysis [21] (Table 1).

Table 1. Summary of advantages and disadvantages of home HD compared to facility HD

	Home	Center/facility
Scheduling	Flexible	Usually fixed, limited ability to change
Hours	Extended hours feasible; including nocturnal	Fixed at 4–5 h per session
Frequency	Any day feasible	Usually locked into 3 times per week
Autonomy	Self care + partner	Limited self-care; often discouraged for "efficiency" reasons
Responsibility	Rests with patient + caretaker	Rests with dialysis staff
Safety	Limited supervision	Fully supervised
Incident management	Rests with patient + caretaker	Managed by staff
Training	6–8 weeks	Nil

Disadvantages

There are some potential disadvantages to home HD. Certainly, the lack of direct supervision is seen by some as a disadvantage, although units that have practiced home care for a long time are far more comfortable with this. The ability to intervene or assist with dialysis technical problems can be troublesome for some, but good training programs can avert much of the danger and anxiety associated with this. Similarly, needling problems often mean a return to the facility for nursing intervention, although home visits by the nurse are increasingly used and can be useful. Inadvertent venous needle dislodgement, especially during nocturnal dialysis, remains an issue for some and there are reported series of these events [22]. Various commercial blood leak detectors are available, but patients are often tired of using them and discard them.

There is a defined technique failure for home HD – quoted as 3–25% per year [23]. This may relate to patients becoming uncomfortable with specific aspects of the self-care of home HD such as problems with needling, fear of dialysis incidents, the burden of responsibility, and the burden placed on care-givers. Several factors may help alleviate this. Providing adequate time for training is important – training is far more than "learning the machine," but is important for the patient's (and care givers) comfort with the process, needling, and their ability to deal with minor events and incidents. If rushed, patients quickly "bail out" when faced with a problem. Previous publications have reported a high technique failure rate in the US, where the NxStage system is widely used. It is our belief that this relates at least in part to attempts to fast track the training rather than problems with the machine itself (for more detail, see [23]).

Patient and caretaker burnout remains a potential problem. It is important to provide opportunities for patients to escape home HD occasionally – this might be via holiday dialysis options such as the Big Red Kidney Bus [24] or simply a return to the home training facility to provide respite dialysis. The provision of respite dialysis is an area that appears under-appreciated. Non-availability of a caretaker can be a problem in some regions, especially jurisdictions that insist on a second person being present for home HD to be undertaken. In established patients, loss of a caretaker (e.g., death, divorce) may then force them back to a facility. In Australia, with the widespread use of nocturnal dialysis, many patients do not dialyze with the assistance of a caretaker, nor is one felt to be necessary.

Barriers

There have been several studies examining the barriers to home HD programs [25–27]. Some of the factors have already been mentioned – fear of needling, concern regarding adopting medical responsibility, and the lack of a caretaker – but a common thread in most studies is education. This relates to both patient and doctor education. The early identification of potential home candidates is essential, with structured education oriented to introducing the concept of home dialysis. Late referral of patients for dialysis is associated with lower rates of home dialysis, and lack of education programs promoting home dialysis is also a factor. Units with active home dialysis programs tend to feed off their own success – staff and patients alike are able to see the benefits of home dialysis first hand and are prepared to promote this.

Staff reticence, uncertainty, and unfamiliarity are clear factors in poor home dialysis uptake. One Australian study revealed medical staff resistance as the major barrier to home HD, even in a country that has a relatively high rate of this therapy [25]. Further, a New Zealand study has revealed the importance placed on medical opinions by patients making a decision around home dialysis [26]. Specific requirements for training in home dialysis (for both PD and home HD) have long been included in the curriculum for advanced training in Nephrology in Australia and New Zealand, and may provide a skill-model for others to follow.

The Future

Given the diverse benefits, home HD should always be considered as a first line treatment for those who require renal replacement therapy. A survey of dialysis patients in Australia showed reluctance to change dialysis modalities once commenced on dialysis [27]. Adopting a home first policy could be implemented to address this. At the least, the option of home dialysis should be considered and discussed with all new patients. The early identification and referral of patients suitable for home dialysis would be a significant step forward. We believe avoiding exposure to facility dialysis in those targeted for home dialysis is also important to avoid the trap of falling into the behavioral pattern of learned helplessness and reliance on center-based dialysis staff.

Technological advances will make home HD easier – easier to learn and easier to implement and monitor. Adopting a strong home HD program is one way to defray the ever-expanding costs of dialysis programs.

Acknowledgement

The authors thank J.W. Agar for useful advice whilst preparing this manuscript.

Disclosure Statement

P.G.K. has participated in advisory boards for Fresenius, Baxter, and Quanta fluid solutions.

References

1 US Renal Data System: Excerpts from the USRDS 2003 annual data report. Chapter 12: international comparisons. Am J Kidney Dis 2003;42(suppl 5):S1–S230.
2 Blagg CR: A brief history of home hemodialysis. Adv Ren Replace Ther 1996;3:99–105.
3 Costas-Lombardia E, Castiel JF: The easy success of the Spanish model for organ transplantation. Artif Organs 2011;35:835–837.
4 Rudge CJ, Buggins E: How to increase organ donation: does opting out have a role? Transplantation 2012;93:141–144.
5 US Renal Data System: USRDS 2015 annual data report. Chapter 13: international comparisons. Am J Kidney Dis 2016;67(suppl 1):S291–S334.
6 ANZDATA Registry: 38th Report, Chapter 6: Home Dialysis. Australia and New Zealand Dialysis and Transplant Registry, Adelaide, Australia. 2016. http://www.anzdata.org.au/anzdata/AnzdataReport/38thReport/c06_anzdata_home_v2.0_20160128_web.pdf (accessed February 1, 2017).
7 Agar JW, Knight RJ, Simmonds RE, Boddington JM, Waldron CM, Somerville CA: Nocturnal haemodialysis: an Australian cost comparison with conventional satellite haemodialysis. Nephrology (Carlton) 2005;10:557–570.

8 Department of Health, Victoria. Funding Policy: Renal Dialysis. https://www2.health.vic.gov.au/hospitals-and-health-services/funding-performance-accountability/pricing-funding-framework/funding-policy/funding-policy-changes/admitted-acute/renal-dialysis (accessed February 1, 2017).
9 Clark WR, Turk JE: The NxStage system one. Semin Dial 2004;17:167–170.
10 Li PK, Chow KM, Van de Luijtgaarden MW, Johnson DW, Jager KJ, Mehrotra R, Naicker S, Pecoits-Filho R, Yu XQ, Lameire N: Changes in the worldwide epidemiology of peritoneal dialysis. Nat Rev Nephrol 2017;13: 90–103.
11 Rajkomar A, Farrington K, Mayer A, Walker D, Blandford A: Patients' and carers' experiences of interacting with home haemodialysis technology: implications for quality and safety. BMC Nephrol 2014;15:195.
12 McCarley P: Patient empowerment and motivational interviewing: engaging patients to self-manage their own care. Nephrol Nurs J 2009;36:409–413.
13 Marshall MR, Polkinghorne KR, Kerr PG, Hawley CM, Agar JW, McDonald SP: Intensive hemodialysis and mortality risk in Australian and New Zealand populations. Am J Kidney Dis 2016;67:617–628.
14 Buchanan C, Mohammed A, Cox E, Kohler K, Canaud B, Taal MW, Selby NM, Francis S, McIntyre CW: Intradialytic cardiac magnetic resonance imaging to assess cardiovascular responses in a short-term trial of hemodiafiltration and hemodialysis. J Am Soc Nephrol 2016;pii:ASN.2016060686.
15 Jeffries HJ, Virk B, Schiller B, Moran J, McIntyre CW: Frequent hemodialysis schedules are associated with reduced levels of dialysis-induced cardiac injury (myocardial stunning). Clin J Am Soc Nephrol 2011;6:1326–1332.
16 Jun M, Jardine MJ, Gray N, Masterson R, Kerr PG, Agar JW, Hawley CM, Van Eps C, Cass A, Gallagher M, Perkovic V: Outcomes of extended-hours hemodialysis performed predominantly at home. Am J Kidney Dis 2013;61:247–253.
17 Liu J, Foley RN: Alternate-day dialysis may be needed for hemodialysis patients. Kidney Int 2012;81:1055–1057.
18 Kerr PG: Dialysis: nocturnal hemodialysis – should we still provide this therapy? Nat Rev Nephrol 2011;7:611–612.
19 Rivara MB, Adams SV, Kuttykrishnan S, Kalantar-Zadeh K, Arah OA, Cheung AK, Katz R, Molnar MZ, Ravel V, Soohoo M, Streja E, Himmelfarb J, Mehrotra R: Extended-hours hemodialysis is associated with lower mortality risk in patients with end-stage renal disease. Kidney Int 2016;90:1312–1320.
20 FHN Trial Group, Chertow GM, Levin NW, Beck GJ, et al: In-center hemodialysis six times per week versus three times per week. N Eng J Med 2010;363:2287–2300.
21 Weinhandl ED, Liu J, Gilbertson DT, Arneson TJ, Collins AJ: Survival in daily home hemodialysis and matched thrice-weekly in-center hemodialysis patients. J Am Soc Nephrol 2012;23:895–904.
22 Pauly RP, Eastwood DO, Marshall MR: Patient safety in home hemodialysis: quality assurance and serious adverse events in the home setting. Hemodial Int 2015;19(suppl 1):S59–S70.
23 Kerr PG, Agar JW: Keeping home dialysis patients at home. Am J Kidney Dis 2016;67: 542–544.
24 Sims J, Bennett PN, Ockerby C, Ludlow M, Fairbairn J, Wilson A, Kerr PG: The effect of holiday haemodialysis treatments on patient mood, adverse symptoms and subjective wellbeing using the big red kidney bus. Nephrology (Carlton) 2017;22:107–113.
25 Ludlow M, George CR, Hawley CM, Mathew TH, Agar JW, Kerr PG, Lauder LA: How Australian nephrologists view home dialysis: results of a national survey. Nephrology (Carlton) 2011;16:446–452.
26 Walker R, Marshall R, Howard K, Morton RL, Marshall MR: "Who matters most?": clinician perspectives of influence and recommendation on home dialysis uptake. Nephrology (Carlton) 2016, Epub ahead of print.
27 Fortnum D, Ludlow R, Morton RL: Renal unit characteristics and patient education practices that predict a high prevalence of home-based dialysis in Australia. Nephrology (Carlton) 2014;19:587–593.

Prof. Peter G. Kerr, MB, BS, PhD, FRACP
Department of Nephrology, Monash Health
246 Clayton Road
Clayton, VIC 3168 (Australia)
E-Mail peter.kerr@monash.edu

La Manna G, Ronco C (eds): Current Perspectives in Kidney Diseases.
Contrib Nephrol. Basel, Karger, 2017, vol 190, pp 156–167 (DOI: 10.1159/000468962)

Treatment of Acute Antibody-Mediated Rejection

Giorgia Comai[a] · Matteo Ravaioli[b] · Olga Baraldi[a] · Vania Cuna[a] · Lorenzo Gasperoni[a] · Giovanni Liviano D'Arcangelo[a] · Maria Cappuccilli[a] · Antonio D. Pinna[b] · Claudio Ronco[c, d] · Gaetano La Manna[a]

[a]Department of Experimental Diagnostic and Specialty Medicine (DIMES), Nephrology, Dialysis and Renal Transplant Unit, and [b]Department of General Surgery and Transplantation, St. Orsola Hospital, University of Bologna, Bologna, [c]Department of Nephrology, Dialysis and Transplantation, and [d]International Renal Research Institute of Vicenza, San Bortolo Hospital, Vicenza, Italy

Abstract

Acute antibody-mediated rejection (AMR) remains an important challenge in the field of kidney transplantation despite the advances in immunosuppressive strategies, and has been recognized as an important cause of allograft dysfunction and graft loss. The treatments of acute AMR are not standardized and poor evidence is currently available on the effectiveness of therapy regimens with combinations of drugs. Standard treatment for acute AMR is based on 3 cornerstones: removal of donor-specific antibodies (DSAs) from the bloodstream, reduction of DSAs synthesis, and inhibition of the interaction between DSAs and human leukocyte antigen antigens on donor's cells. In this paper, we discuss the mechanism of action of the principal drugs currently in use, the main published studies on this topic, and the future insights related to the treatment of acute AMR.

Introduction

Renal transplant represents the treatment of choice for patients with end-stage renal disease (ESRD), often leading to considerable amelioration or resolution of the most common dialysis-related complications [1–3], although transplant outcomes are significantly influenced by several immunological, non-immunological, genetic, and donor-related factors [4–6].

In the last decades, the improved awareness of the immune system, together with the growing advances in immunosuppressive strategies, led to a significant reduction in the incidence of acute rejection. In particular, the rate of acute rejection occurring within the first year after transplantation decreased from 50% in the 1990s to 7.8% in 2014 [7].

Despite this marked reduction, about 1–10% of kidney transplant recipients experience acute antibody-mediated rejection (AMR), which frequently co-exists with acute cellular rejection, and is largely related to patient's pre-transplant immunization status and to the immunosuppressive regimens administered [8].

AMR, also known as acute humoral rejection, is a destructive immune reaction due to the presence of alloantibodies directed against the transplanted organ, that can be specific either to organ donor human leukocyte antigen (HLA; donor-specific antibodies, DSAs) or to non-HLA anti-endothelial antigens (non-DSAs), namely angiotensin II type 1 receptor antibodies, major histocompatibility complex class-I-related chain A antibodies (MICA), C-terminal laminin G antibodies (anti LG3), anti-fibronectin and anti-collagen type 4 autoantibodies, and others [9, 10]. Independent of the type of antibodies, allograft damage results from the interaction between antibodies and graft endothelial cells. It promotes the activation of classical complement pathways, lymphocyte chemotaxis, osmotic cell lysis, release of von Willebrand factor, platelet activation, and finally endothelial injury [8, 9]. It is also known a C4d-negative form of AMR, feasibly due to a complement independent mechanism which occurs through the binding between the antibodies Fc region to natural killer (NK) cells receptor with the release of cytotoxic cytokines [9].

AMR is more commonly observed within 3 months post-transplant, but it might also occur later, especially in case of low immunosuppression or non-adherence. Generally, the prognosis of AMR is worse than that of acute cellular rejection, and 1-year graft loss rate following AMR ranges from 15 to 20%. AMR is less responsive to the conventional anti-rejection therapy, and is an important predictor of chronic allograft dysfunction and graft loss, so AMR prevention and treatment are key targets to improve long-term kidney transplant survival [7, 9].

The diagnostic criteria for acute/active AMR, as suggested by the last Banff kidney meeting report, consist of: (1) histologic evidence of acute tissue injury including microvascular inflammation, intimal or transmural arteritis, acute thrombotic microangiopathy (TMA) in the absence of any other cause and acute tubular injury, and in the absence of any other apparent cause; (2) evidence of current/recent antibody interaction with vascular endothelium, including linear C4d staining in peritubular capillaries, at least moderate microvascular inflammation, increased expression of gene transcripts in the biopsy tissue indicative of endothelial injury; (3) serological evidence of DSAs (HLA or other antigens) [11].

The 2009 Kidney Disease Improving Global Outcomes clinical practice guidelines for the care of kidney transplant recipients recommend the standard treatment for acute AMR to be based on 3 cornerstones: (i) removal of DSAs from the bloodstream (plasma-exchange), (ii) reduction of DSA synthesis (lymphocytes-depleting antibodies, anti-CD20 antibodies, and intravenous immunoglobulin [IVIg] therapy), and (iii) inhibition of the interaction between DSAs and HLA antigens on donor's cells (IVIg therapy) [12]. High doses of steroids are a non-specific option for AMR treatment and only 50% of the patients respond with a graft survival ranging from 15 to 65% [8].

In this paper, we present a case report and our case series of acute AMR, followed by a literature review of AMR therapy.

Case Report

A 65-year-old female with ESRD due to autosomal dominant polycystic kidney disease, received a kidney transplant from a 66-year-old deceased donor. The patient's panel reactive antibody level was 0%, and the transplant was executed with 5 HLA-mismatches (2 class A, 2 class B, 1 class DR), negative T- and B-cell complement-dependent cytotoxicity crossmatch, and a cold ischemia interval of 8 h. Induction therapy was based on basiliximab, steroids, tacrolimus 0.2 mg/kg, mycophenolic acid 720 mg/day.

The patient experienced delayed graft function (DGF) during the first week post transplant, with elevated serum creatinine levels (6.5 mg/dL), normal hemoglobin (11.1 g/dL), low platelet count (132,000/μL), and tacrolimus level within the normal range (8 ng/mL). The patient had no fever or other infection signs and a normal graft ultrasound (11 cm in length, normal echogenicity, no hydronephrosis, and regular arterial flow). Because of the persistence of DGF, on day 10 we performed a kidney biopsy and measured DSAs.

The histological analysis indicated the presence of acute tissue injury characterized by microvascular inflammation (glomerulitis g 1 and glomerular thrombi), intimal arteritis (v 2), acute tubular injury; evidence of antibody interaction with the endothelium characterized by peritubular capillaritis (ptc 2), C4d stain negative, and electron microscopy confirmed all the features displayed by light microscopy (Fig. 1, left). The serological detection of DSAs revealed a median mean fluorescence intensity of 7100 for anti-DR11 antibodies. The presence of all these features suggested a diagnosis of acute AMR according to the 2015 updated Banff classification [11].

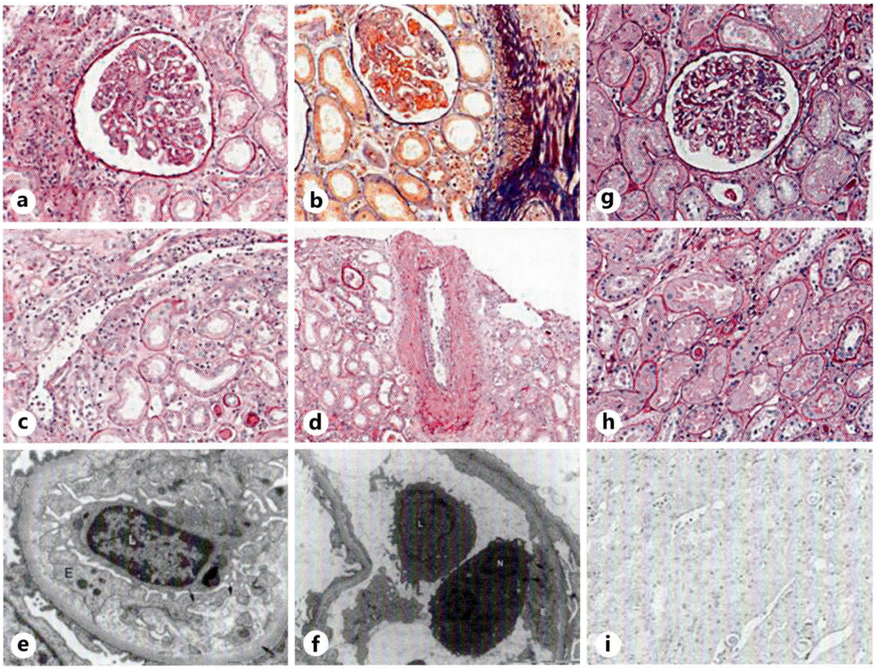

Fig. 1. First biopsy (left). Light microscopy: **a** glomerulitis 1+ and mesangiolysis; **b** glomerular thrombi; **c** peritubular capillaries: dilated with neutrophils as peritubular capillaritis 2+, interstitium: edema and sparse infiltrate, tubules: acute tubular necrosis; **d** vascular inflammation intimal arteritis (v 2). Electron microscopy: **e** partial occlusion of a glomerular capillary by leukocyte, lymphocyte; (L) and neutrophil (N) infiltration; note close adherence of neutrophil (N) with endothelial cell (arrows). Scale bar = 5 µm; **f** a glomerular capillary completely occluded by lymphocyte (L) showing endothelial cell (E) swelling, subendothelial electron-lucent widening and early basal lamina duplication (arrows). Scale bar = 2 µm. Second biopsy (right). Light microscopy: **g** glomerulitis 1+; **h** peritubular capillaritis 1+, interstitium normal; tubules with some casts; no vascular inflammation; **i** immunohistochemistry C4d negative.

The patient received a therapy based on antithymocyte globulin (ATG) 1 mg/kg for 3 days, followed by 3 plasmapheresis (PP) treatments, and 3 IVIg infusions 0.1 g/kg. We observed an improvement in urine output and serum creatinine that dropped to 2 mg/dL, and a negativization of DSAs, so the patient was discharged after 25 days of hospital stay.

Five months after transplant, serum creatinine was 1.7 mg/dL, tacrolimus level was 7.5 ng/mL, and urine protein test was negative. No DSAs were detected. We performed a second biopsy that showed the absence of glomerular thrombi, glomerulitis (g 1), normal tubules, no interstitium infiltrates, no arteritis, negative C4d (Fig. 1, right). At the last follow-up (18 months), serum creatinine was 1.7 mg/dL, urine protein was negative, and DSAs were negative.

Treatment of AMR

At the best of our knowledge, there are no randomized controlled trials (RCTs) on AMR treatment with adequate statistical power to compare safety and efficacy of different treatment options. Actually KDIGO suggests the use of one or more previously described strategies, graded 2C, according to the GRADE system to rate the strength of evidence and the strength of recommendations [12].

In general, most of the treatments consisted of a combination of apheresis with 2 or more drugs, namely IVIg, ATGs, rituximab, bortezomib, and more recently eculizumab without differences in terms of efficacy between all the strategies.

Table 1 reports a list of recently published controlled trials and retrospective studies investigating different drug combinations, and Table 2 describes our series of AMR treated with different approaches.

We describe below the mechanism of action and the main applications of some drug combinations targeted to the different pathogenic mechanisms underlying AMR.

The goal of PP and other immunoabsorption techniques is the removal of alloantibodies, in particular DSAs, from the circulation. They have been used for the treatment of AMR since the 1970s, always in combination with anti-rejection drugs. The number and type of apheresis treatment can be variable among the published studies, and poor evidence from solid RCTs is currently available to confirm the effectiveness of PP in the therapy of acute AMR [13–15]. However, despite this and other limitations, apheresis remains a commonly used approach for treating AMR [16].

IVIg exert their effect on AMR by reducing the activity of both humoral and cellular immunity, as well as by inhibiting the interaction between DSAs and donor's antigens. To express all their therapeutic effect, IVIg need the binding of Fc fragment with Fc receptors present on immune system cells. Different IgG regions may target DSAs or other autoantibodies, cytokines, complement fragments, determine inhibition of dendritic cells and macrophages and apoptosis of plasma cells and could determine the destruction of immune cells through antibody-dependent cellular cytotoxicity [17]. Moreover, recent studies showed that IgG may induce dendritic cells to produce combinations of cytokines able to stimulate T regulatory cells and inhibit Th17 cell polarization [18]. However, despite the theoretical assumptions, experimental data on the clinical efficacy of IVIg in acute AMR is scarce, and therapeutic protocols and dosages are not standardized (IVIg doses in the literature vary between 0.1 and 2 g/kg and vary depending on PP employment), as summarized in Table 2 [14, 15, 19].

Table 1. Recent published controlled trials and retrospective studies investigating different drug combinations in AMR

Controlled trials	Sample size	Treatment	Efficacy
Böhmig et al. [13], 2007	10 (5 treatment arm, 5 control arm)	IA (daily for 3 days, then every 3 days up to 6 weeks)	Yes
Lefaucheur et al. [26], 2009	24 (12 treatment arm, 12 control arm)	All treated with methylprednisolone 500 mg/day for 3 days Treatment arm: PP (4 sessions), IVIg 0.1 g/kg after PP then 2 g/kg every 3 weeks × 4 doses, rituximab 375 mg/m^2 (2 weekly doses) Control arm: IVIg 2 g/kg (4 doses)	Yes
Kaposztas et al. [27], 2009	54 (26 treatment arm, 28 control arm)	Treatment arm: PE (5 daily followed by 2 days rest; thereafter, thrice weekly for 2–3 weeks), rituximab 375 mg/m^2 after each PE course Control arm: PE (same dose as group A)	Yes
Sautenet et al. [28], 2016	40 (21 treatment arm, 19 control arm)	Treatment arm: PP (at least 3), IVIg (0.1 g/kg after PP until day 4 then 1 g/kg/day on days 5 and 6), methylprednisolone 500 mg/day (for 3 days), rituximab (375 mg/m^2 on day 5) Control arm: PP, IVIg, methylprednisolone (same doses as treatment arm), placebo	No

Retrospective studies	Sample size	Treatment	Follow up	Outcome
Pretagostini et al. [14], 2012	30	IA (mean 7 sessions) vs. DFPP (mean 7 sessions) All treated with IVIg (400 mg/kg/die)	1 year	No difference in allograft survival and DSA removal rate
Gubensek et al. [15], 2016	43	PP (daily for 3 days then every other day) or IA (twice for thrice a week) mean 8 sessions + IVIg (100 mg/kg after PP)	5 years	Immediate allograft function improvement but progressive decline during follow up
Ruangkanchanasetr et al. [19], 2015	13	PP (mean 8, daily), IVIg (2±1 g/kg after PP); stop after control biopsy, if no response after 3 PP, rituximab or bortezomib	8.7 months	Allograft function improvement in most patients, 1 patient resumed hemodialysis
Zheng et al. [23], 2014	5	ATG 75 mg/day (for 4–8 days), double filtration PP (4 sessions), IVIg 50 g/day (for 2 days), rituximab 375 mg/m^2 (single dose)		Allograft function improvement only in 1 patient with mild acute AMR
De Sousa-Amorim et al. [29], 2016	7	Rituximab 400 mg (2 doses), PP (mean 5 sessions), bortezomib 1.3 mg/m^2 after PP, IVIg 0.1 g/kg after PP	22 months	At the end of follow-up, 3 grafts were lost and 3 remained functioning (all cases were severe forms of acute AMR)
Kizilbash et al. [30], 2017	33 (pediatric)	IVIg 2 g/kg (2 doses), PP (5 sessions), rituximab 366 mg/m^2 (2 doses), bortezomib 5.2 mg/m^2 (3–8 doses)	15 months	Renal function improved or stabilized in most patients
Tran et al. [32], 2016	1	Pulse steroid therapy (3 doses), ATG (2 doses), PP (3 sessions) IVIg (4 doses), rituximab (4 doses) + eculizumab 3 doses (1,200, 900 and 900 mg)	1 year	Progressive allograft function decline (signs of TMA)
Burbach et al. [33], 2014	2	Eculizumab (900 mg per week for 4 weeks and then 1,200 mg per 2 weeks), PP, IVIg, rituximab	2 years	No improvement in allograft function

AMR, antibody-mediated rejection; ATG, antithymocyte globulin; DFPP, double filtration plasmapheresis; IA, immunoadsorption; IVIg, intravenous immunoglobulin therapy; PE, plasma exchange; PP, plasmapheresis; TMA, thrombotic microangiopathy.

Table 2. Case series of AMR treated with different approaches at Renal Transplant Center, Bologna University

Treatment	DSA	Biopsy	Results
Steroids 500 mg for 3 days followed by IVIg 2 g/kg (5 doses)	DQ2 (MFI 5900)	g 1, ptc 1, C4d 2	Immediate allograft function improvement but progressive decline, and in 6 months patient resumed dialysis
Steroids 500 mg for 3 days, plasmapheresis (2 sessions) followed by IVIg 0.1 g/kg after PP, rituximab 375 mg/m^2, eculizumab 900 mg	DQ7 (MFI 10340)	g 2, ptc 2, C4d 1, t 1, i 1, acute tubular necrosis, thrombotic microangiopathy	Allograft function improvement, serum creatinine 1.8 mg/dL; follow-up 9 months
Antithymocyte globulin 1 mg/kg × 5 days followed by 3 plasmapheresis and intravenous immune globulin infusion 0.1 g/kg after PP, rituximab 375 mg/m^2	B 7 (MFI 15000)	g 1, ptc 2, C4d 2	Allograft function improvement, serum creatinine 1.1 mg/dL; follow-up 6 years
Antithymocyte globulin 1 mg/kg × 3 days followed by 3 plasmapheresis and intravenous immune globulin infusion 0.1 g/kg after PP	B35 (MFI 22000)	ptc 2, C4d 2, ti 1	Allograft function improvement, serum creatinine 1.4 mg/dL; follow-up 1 year
Antithymocyte globulin 1 mg/kg × 3 days followed by 3 plasmapheresis and intravenous immune globulin infusion 0.1 g/kg after plasmapheresis	DR11 (MFI 7100)	g 1, ptc 2, v 2 acute tubular necrosis	Allograft function improvement, serum creatinine 1.4 mg/dL; follow-up 1 year
Antithymocyte globulin 1 mg/kg × 3 days followed by 6 plasmapheresis and intravenous immune globulin infusion 0.1 g/kg after plasmapheresis	DR11 (MFI 7100)	g 2, ptc 1, C4d 1, t1, i1	Allograft function partial improvement, serum creatinine 3 mg/dL; follow-up 1 month

g, glomerulitis; ptc, peritubular capillaritis; i, interstitial inflammation; t, tubulitis; v, intimal arteritis.

Although the role of ATGs as induction therapy in kidney transplantation is well established, their application in the treatment of AMR is not supported by strong evidence yet. Rabbit ATGs (rATGs), the most utilized worldwide, consist of polyclonal IgG fractions purified from sera of rabbit immunized with human lymphocytes, and they interact with a variety of antigens on immune and non-immune cell types, inducing apoptosis of peripheral T-cells, B-cells, plasma cells, NK cells, and modulate leukocyte/endothelium interactions [20, 21]. The mechanisms of action are antibody-dependent cell cytotoxicity (the main mechanism triggered by the fixation of the polyclonal antibodies on antigens), complement-dependent cell cytotoxicity (dose dependent), opsonizations, and cell death pathways [20]. Moreover, rATGs have also been proven to suppress donor-reactive memory T-cells, and to create a preferential reconstitution of T-regulatory cells (Tregs), a cellular subset involved in the immunomodulation and control of peripheral tolerance in organ transplant recipients [21, 22]. In recent studies, ATGs have been used to treat mixed rejection (T-cell and antibody-mediated), suggesting a potential efficacy in combined treatment with PP [20, 23]. However, the choice of ATGs as first line therapy is not recommended, because of its poor in vivo effect on B cells [20].

Among the agents with a feasible efficacy on B-cells, rituximab is a chimeric antibody directed against the cell-surface antigen CD20, commonly expressed during B-cell development, except in the very early stages and plasma cells. Rituximab elicits the depletion of CD20+ B-cells, followed by a gradual reconstitution of peripheral blood B cells over the subsequent months and a reduction in spleen and lymph nodes [24]. The efficacy of a combined therapy with rituximab, PP and IVIg, has been demonstrated to prevent AMR, as in the setting of AB0-incompatible kidney transplantation, probably because it determines immature B-cell depletion prior to the contact between DSAs and donor cells [25]. Moreover, rituximab does not seem to have a direct effect on antibody-producing plasma cells, which could explain the modest results in the treatment of established acute AMR [24]. There is some evidence to support that rituximab could have some beneficial effects in the treatment of acute AMR in addition to standard therapy [25–27]. The most recent RCT analyzing the effect of rituximab in acute AMR is a French multicenter, prospective, randomized placebo-controlled clinical trial published in 2016. In this RCT, 38 patients with biopsy-proven acute AMR diagnosed in the first year after transplantation were randomly assigned to receive rituximab (375 mg/m^2 single dose) or placebo at day 5. All patients were treated with PP, IVIG, and corticosteroids. After 1 year of follow-up, this RCT failed to demonstrate an additional effect of rituximab in patients receiving standard treatments for acute AMR [27]. However, it is likely that the results might have been probably underpowered by the low number of participants and the short follow-up that limited the possibility to demonstrate a clear beneficial effect of the drug in the long-term, also in view of rituximab pharmacodynamics.

Bortezomib, a modified dipeptidyl boronic acid, is a reversible inhibitor of the 26S proteasome, with the ability to trigger apoptosis of bone marrow-derived plasma cells. In a literature review of 2012, Roberts et al. [16] concluded that only controlled but not randomized studies supported the usefulness of bortezomib in the therapy of acute AMR, in the absence of solid evidences. After 2012, 6 works focused on the use of bortezomib in acute AMR have been published, 3 of them are retrospective studies and the other 3 are case reports (almost all of the studies performed on pediatric populations; Table 2) [29, 30]. In spite of the absence of RCTs in the literature, bortezomib seems to emerge as a promising option for acute AMR, although strong evidences are still needed.

Given the involvement in AMR of complement activation as a powerful mechanism leading to graft injury and damage [9], eculizumab, a high-affinity monoclonal, humanized antibody directed against complement component C5, has been proposed for AMR treatment [31]. Eculizumab inhibits the formation of the C5–C9 complex or membrane attack complex and the production of the anaphylatoxin C5a, thus preventing complement-mediated tissue injury. The activation of the complement cascade is a key pathway in acute AMR pathogenesis, and C4d deposition in allograft capillaries is considered a hallmark criterion for the diagnosis of AMR [32]. To date, there are 13 case reports describing the use of eculizumab in the treatment of acute AMR resistant to standard treatments [33, 34]. In the great majority of these cases, eculizumab has been effective in reversing the rejection episode and improving the graft function. However, cases of C4d-negative acute AMR did not respond well to eculizumab, suggesting alternative mechanisms involved, other than complement activation [32].

Complement cascade blockage at the level of C5, however, does not protect against the formation of pro-inflammatory molecules, generated during the early phases of complement activation (e.g., C3a) [32]. Taking this drawback into account, a new approach to counteract complement activity might be represented by C1 inhibition. Recent studies demonstrated that the capacity of anti-HLA antibodies to bind complement fraction C1q, is associated with poor outcome and severe forms of antibody-mediated injury [34]. This information provides a further rationale for the development of complement-targeting therapies using C1 inhibitors (C1-INHs) [35]. C1-INH, which is a member of the serpin family of protease inhibitors, blocks C1-complex activity by the inactivation of C1r and C1s. A phase 1/2 study on 20 patients evaluating the role of C1-INH in the treatment of acute AMR in kidney transplantation showed a favorable safety profile and a potential benefit by reducing the capacity of anti-HLA antibodies to bind C1q [36].

Considerations and Conclusions

In summary, most of the studies cited are single cases yielding a patchy picture with conflicting results, due to wide heterogeneity among clinical, immunological, and histological presentations of acute AMR and in the clinical management. In our opinion, there are still 4 "puzzling" points in the AMR setting that could explain the difficulties in the treatment and approaches: (1) not all DSA are equally pathogenic; (2) in some cases of AMR, C4d might be negative, and however, even if positive alone, is not diagnostic; (3) the emerging role of non-

HLA antibodies in AMR pathogenesis; (4) the relationship between acute and chronic AMR.

A very important point is that B-cells play a crucial role in the pathogenesis of AMR, because they are precursors of plasma cells which produce DSAs that can bind to the allograft vasculature and activate complement, thus determining tissue injury, but they can also participate in antibody-independent immune responses through antigen presentation [8]. In addition, AMR can also be associated with a T-cell-mediated component, essential for the production of DSAs IgG antibodies [9]. NK transcripts and NK cells have been documented in kidney biopsies from recipients with DSAs and acute AMR, suggesting a pathogenic role of this cell subset. Complement system has a paramount importance in acute AMR pathogenesis. Complement activation, confirmed by capillary deposition of complement split-product C4d, has been implicated in acute AMR-induced graft damage and, more generally, with acute or chronic micro-circulation inflammation, and injury [10]. However, emerging evidence has shown that acute AMR does not necessarily associate with detectable capillary C4d deposits and that C4d-negativity on graft biopsy does not exclude a causative role of complement activation, as reported in Banff 2015 Classification [11].

Taking into account all these mechanisms, we report our experience in AMR cases as summarized in Table 2. We tried different strategies based on histological lesion. The use of ATGs, in association with other therapies, represents a reliable option for the treatment of established acute AMR, especially when T-cell mediated rejection coexists. Finally, eculizumab or other complement inhibitors appear to be a good choice in the presence of signs of TMA.

References

1 Meier-Kriesche HU, Schold JD, Srinivas TR, Reed A, Kaplan B: Kidney transplantation halts cardiovascular disease progression in patients with end-stage renal disease. Am J Transplant 2004;4:1662–1668.

2 Cianciolo G, Colí L, La Manna G, et al: Is beta2-microglobulin-related amyloidosis of hemodialysis patients a multifactorial disease? A new pathogenetic approach. Int J Artif Organs 2007;30:864–878.

3 Cianciolo G, La Manna G, Colì L, et al: 5-methyltetrahydrofolate administration is associated with prolonged survival and reduced inflammation in ESRD patients. Am J Nephrol 2008;28:941–948.

4 La Manna G, Lanci N, Della Bella E, et al: Reduction of oxidative damage reflects a better kidney transplantation outcome. Am J Nephrol 2011;34:496–504.

5 La Manna G, Cappuccilli ML, Cianciolo G, et al: Cardiovascular disease in kidney transplant recipients: the prognostic value of inflammatory cytokine genotypes. Transplantation 2010;89:1001–1008.

6 La Manna G, Comai G, Cappuccilli ML, et al: Prediction of three-year outcome of renal transplantation from optimal donors versus expanded criteria donors. Am J Nephrol 2013;37:158–166.

7 Hart A, Smith JM, Skeans MA, et al: OPTN/SRTR 2015 annual data report: kidney. Am J Transplant 2017;17(suppl 1):21–116.
8 Chehade H, Pascual M: The challenge of acute antibody-mediated rejection in kidney transplantation. Transplantation 2016;100: 264–265.
9 Davis S, Cooper JE: Acute antibody-mediated rejection in kidney transplant recipients. Transplant Rev (Orlando) 2017;31:47–54.
10 Cardinal H, Dieudè M, Hebert MJ: The emerging importance of non-HLA autoantibodies in kidney transplant complications. J Am Soc Nephrol 2017;28:400–406.
11 Loupy A, Haas M, Solez K, et al: The Banff 2015 kidney meeting report: current challenges in rejection classification and prospects for adopting molecular pathology. Am J Transplant 2017;17:28–41.
12 Kidney Disease: Improving Global Outcomes (KDIGO) Transplant Work Group: KDIGO clinical practice guideline for the care of kidney transplant recipients. Am J Transplant 2009;9(suppl 3):S1–S155.
13 Böhmig GA, Wahrmann M, Regele H, et al: Immunoadsorption in severe C4d-positive acute kidney allograft rejection: a randomized controlled trial. Am J Transplant 2007;7: 117–121.
14 Pretagostini R, Poli L, Gozzer M, et al: Plasmapheresis, photopheresis, and endovenous immunoglobuline in acute antibody-mediated rejection in kidney transplantation. Transplant Proc 2015;47:2142–2144.
15 Gubensek J, Buturovic-Ponikvar J, Kandus A, et al: Treatment of antibody-mediated rejection after kidney transplantation – 10 years' experience with apheresis at a single center. Ther Apher Dial 2016;20:240–245.
16 Roberts DM, Jiang SH, Chadban SJ: The treatment of acute antibody-mediated rejection in kidney transplant recipients-a systematic review. Transplantation 2012;94:775–783.
17 Tedla FM, Roche-Recinos A, Brar A: Intravenous immunoglobulin in kidney transplantation. Curr Opin Organ Transplant 2015;20: 630–637.
18 Kessel A, Ammuri H, Peri R, Pavlotzky ER, Blank M, Shoenfeld Y, Toubi E: Intravenous immunoglobulin therapy affects T regulatory cells by increasing their suppressive function. J Immunol 2007;179:5571–5575.
19 Ruangkanchanasetr P, Satirapoj B, Termmathurapoj S, et al: Intensive plasmapheresis and intravenous immunoglobulin for treatment of antibody-mediated rejection after kidney transplant. Exp Clin Transplant 2014; 4:328–333.
20 Bamoulid J, Staeck O, Crepin T, et al: Antithymocyte globulins in kidney transplantation: focus on current indications and long-term immunological side effects. Nephrol Dial Transplant 2016;pii:gfw368.
21 Zand MS, Vo T, Huggins J, et al: Polyclonal rabbit antithymocyte globulin triggers B-cell and plasma cell apoptosis by multiple pathways. Transplantation 2005;79:1507–1515.
22 La Manna G, Ghinatti G, Tazzari PL, et al: Neutrophil gelatinase-associated lipocalin increases HLA-G(+)/FoxP3(+) T-regulatory cell population in an in vitro model of PBMC. PLoS One 2014;9:e89497.
23 Zheng J, Xue W, Qing X, et al: Treatment of biopsy-proven acute antibody-mediated rejection using thymoglobulin (ATG) monotherapy and a combination of rituximab, intravenous immunoglobulin, and plasmapheresis: lesson learned from primary experience. Clin Transpl 2014:223–230.
24 Barnett AN, Hadjianastassiou VG, Mamode N: Rituximab in renal transplantation. Transpl Int 2013;26:563–575.
25 Zarkhin V, Li L, Kambham N, Sigdel T, Salvatierra O, Sarwal MM: A randomized, prospective trial of rituximab for acute rejection in pediatric renal transplantation. Am J Transplant 2008;8:2607–2617.
26 Lefaucheur C, Nochy D, Andrade J, et al: Comparison of combination Plasmapheresis/IVIg/anti-CD20 versus high-dose IVIg in the treatment of antibody-mediated rejection. Am J Transplant 2009;9:1099–1107.
27 Kaposztas Z, Podder H, Mauiyyedi S, et al: Impact of rituximab therapy for treatment of acute humoral rejection. Clin Transplant 2009;23:63–73.
28 Sautenet B, Blancho G, Büchler M, et al: One-year results of the effects of rituximab on acute antibody-mediated rejection in renal transplantation: RITUX ERAH, a multicenter double-blind randomized placebo-controlled trial. Transplantation 2016;100:391–399.

29 De Sousa-Amorim E, Revuelta I, Diekmann F, et al: Bortezomib for refractory acute antibody-mediated rejection in kidney transplant recipients: a single-centre case series. Nephrology (Carlton) 2016;21:700–704.
30 Kizilbash S, Claes D, Ashoor I, et al: Bortezomib in the treatment of antibody-mediated rejection in pediatric kidney transplant recipients: a multicenter Midwest Pediatric Nephrology Consortium study. Pediatr Transplant 2017, Epub ahead of print.
31 Stegall MD, Chedid MF, Cornell LD: The role of complement in antibody-mediated rejection in kidney transplantation. Nat Rev Nephrol 2012;8:670–678.
32 Tran D, Boucher A, Collette S, Payette A, Royal V, Senécal L: Eculizumab for the treatment of severe antibody-mediated rejection: a case report and review of the literature. Case Rep Transplant 2016;2016:9874261.
33 Burbach M, Suberbielle C, Brochériou I, et al: Report of the inefficacy of eculizumab in two cases of severe antibody-mediated rejection of renal grafts. Transplantation 2014;98: 1056–1059.
34 Moktefi A, Parisot J, Desvaux D, et al: C1q binding is not an independent risk factor for kidney allograft loss after an acute antibody-mediated rejection episode: a retrospective cohort study. Transpl Int 2017;30:277–287.
35 Locke JE, Magro CM, Singer AL, et al: The use of antibody to complement protein C5 for salvage treatment of severe antibody-mediated rejection. Am J Transplant 2009;9: 231–235.
36 Montgomery RA, Orandi BJ, Racusen L, et al: Plasma-derived C1 esterase inhibitor for acute antibody-mediated rejection following kidney transplantation: results of a randomized double-blind placebo-controlled pilot study. Am J Transplant 2016, Epub ahead of print.

Gaetano La Manna, MD, PhD
Department of Experimental, Diagnostic and Specialty Medicine (DIMES)
Nephrology, Dialysis and Transplantation Unit
St. Orsola Hospital, University of Bologna
Via G. Massarenti 9 (Pad. 15), IT–40138 Bologna (Italy)
E-Mail gaetano.lamanna@unibo.it

La Manna G, Ronco C (eds): Current Perspectives in Kidney Diseases.
Contrib Nephrol. Basel, Karger, 2017, vol 190, pp 168–180 (DOI: 10.1159/000468963)

Chronic Allograft Injury: An Overview of Pathogenesis and Treatment Strategies

Hector M. Madariaga[a] · Leonardo V. Riella[b]

[a]Department of Medicine, Good Samaritan Medical Center, Brockton, MA, and [b]Renal Division, Brigham and Women's Hospital, Harvard Medical School, Boston, MA, USA

Abstract

Chronic allograft injury (CAI) remains one of the biggest challenges in transplantation as it directly affects the long-term allograft survival, an area in which the scientific and transplantation communities have struggled to improve outcomes. Therefore, understanding the mechanisms of CAI is paramount to implement preventive measures and novel therapeutic options. In this review, we will address the latest evidence on CAI, potential etiologies and risk factors including non-adherence, antibody-mediated rejection, recurrence of glomerular disease, BK nephropathy, and calcineurin nephrotoxicity.

Introduction

The term chronic allograft nephropathy is no longer used to describe kidney allografts with chronic lesions since it fails to capture the etiology of chronic damage and falsely creates a single "diagnosis" without taking into account the heterogeneity of the underlying etiologies. As per the revised Banff 2005 classification system, this condition was renamed chronic allograft injury (CAI), attempting to reinforce the search for the cause of injury [1]. This is a major challenge in transplantation, as short-term outcomes improved with reduced rejection rates while long-term outcomes remained poor. By 10 years post-transplant, half of the allografts are lost due to non-immunological and/or immunological factors. In addition, death-censored rates for standard criteria donor kidneys (3–5%/year) and for living donor kidneys (2–3%/year) have not

Table 1. Causes of chronic allograft injury

Non-immunological factors	Immunological factors
Recipient's cardiovascular risks (hypertension, hyperlipidemia, diabetes mellitus)	Donor-specific antibodies prior to transplant or de novo
Prolonged cold and warm ischemia time	Acute rejection in the first year and recurrent episodes
CNI nephrotoxicity	Antibody-mediated rejection
Poor quality of grafts from donors	Viral (BK and CMV) and bacterial infections and recurrent infections
Urological issues	Recurrent glomerular disease
Renal artery stenosis	
Significant size mismatch between recipient and donor kidney	

changed in the last 3 decades [2], and death with a functioning graft has been reported to be one of the most common causes of graft loss post-transplant. In a single center study in which protocol biopsies were performed routinely, glomerular disease, interstitial fibrosis/tubular atrophy (IFTA), and acute rejection were the most common etiologies of CAI [3]. In patients with IFTA, prior protocol biopsies and medical history were assessed revealing potential mechanism of injury including cellular and/or antibody-mediated rejection and BK nephropathy. Other less common causes are related to poor quality donor kidneys, ureteral complications and infections such as recurrent pyelonephritis and calcineurin inhibitor (CNI) toxicity. In the subsequent sections, we will describe in more depth the immunological and non-immunological factors that may contribute to CAI (Table 1).

Non-Immunological Factors

CNI Nephrotoxicity

Kidney transplant recipients treated with CNI (tacrolimus or cyclosporine) are at risk of nephrotoxicity. The mechanism by which CNI causes injury is through vasoconstriction, tubular and vascular injury, and less commonly through thrombotic microangiopathy. In the long-term, kidneys exposed to CNI develop IFTA, glomerulosclerosis, and arteriolar hyalinosis. The ELITE-Symphony trial, which compared standard and low-dose cyclosporine, low-dose tacrolimus, and low-dose sirolimus-based therapy revealed that low- to moderate-dose tacrolimus (target 3–7 ng/mL, median 7 ng/mL), in combination with mycophenolate and steroids, was associated with lower fibrosis, lower rejection rate, and improved outcomes at 1 year post-transplant [4]. Further analysis after 3 years post-transplant, confirmed the benefit of lower tacrolimus levels with lower rejection rates and improved graft survival [5].

It is evident that elevated levels of CNI are associated with nephrotoxicity as we observe in non-kidney transplants, in which the higher the CNI exposure, the greater the rate of chronic kidney disease. Snanoudj et al. [6], compared over the course of 10 years, 2 groups of kidney transplant recipients, one receiving CNI and other group who did not. Late biopsies in both groups, revealed arteriolar hyalinosis in 92% of patients on CNI and in 65% of patients who were not on CNI. This study questioned the association of allograft chronic with CNI exclusively. Current protocols use lower CNI through levels, but it is important to consider that tacrolimus remains the cornerstone of immunosuppression in kidney transplantation. Indeed, CNI withdrawal studies have shown a high rate of de novo donor-specific antibodies (DSA) development and rejection, indicating the importance of this agent as a component of the immunosuppressive regimen [7].

Poor Quality Donor Grafts

Kidneys from donors after cardiac death, and who are elderly, diabetic or with hypertension may lead to shorter allograft survival as these allografts may be sensitive to new insults and have a reduced residual renal functional capacity. Furthermore, certain genetic risk factors from the donor may also influence graft outcomes. For example, the presence of 2 high-risk alleles for APOL1 is associated with worse graft survival [8, 9], and recent preclinical study has confirmed this association indicating that high-risk APOL1 produced by the kidney is toxic to podocytes and may lead to glomerular injury [10]. Even the incompatibility of donor and recipient weight (kidney weight and recipient weight ratio <2.3 g/kg) is associated with lower graft survival [11], due to glomerular hyperfiltration and hypertrophy leading to glomerular injury. Lastly, the duration of cold ischemia time is also associated with CAI due to a combination of direct kidney injury and higher risk of rejection induced by ischemic injury and innate immune activation [12].

Uncontrolled Hypertension

Elevated blood pressures result in poor allograft survival [13]. In up to 5% of kidney recipients, renal artery stenosis can be the etiology of persistent uncontrolled hypertension. It is usually secondary to stenosis of the anastomotic site or presence of an atherosclerotic plaque in the transplant artery. Diagnosis requires a Doppler ultrasound, MRI, or arteriography (gold-standard) and treatment involves angioplasty. Recognition of renal artery stenosis is important based on possibility of intervening and correcting it.

Other Causes
Hyperlipidemia has been associated with CAI. Guijarro et al. [14] reported that post-transplant hypertriglyceridemia, but not hypercholesterolemia, may be associated with CAI. However the exact mechanism is unclear and further evidence is required to confirm this. Some preclinical studies suggest that hyperlipidemia may trigger rejection possibly by promoting anti-donor Th17-cell response [15, 16]. Ureteral obstruction such as stenosis has also been reported to cause CAI and is frequently a consequence of ureter ischemia. A graft ultrasound is used for diagnosis to identify hydronephrosis.

Immunological Causes

Multiple immunological factors are involved in the etiology of CAI. Allo-MHC molecules, normally present in the donor kidney, are targets for a chronic immunologically mediated injury. The rate of acute rejection is about 5% per year after the first year post-transplant, and chronic rejection is the most common type of rejection after transplant. Chronic rejection entails an immune response to the allograft, characterized by cellular and/or antibody-mediated injury. The role of DSA has been increasingly recognized as a cause of late graft dysfunction and there are several risk factors involved in the development of DSA, such as under immunosuppression and non-adherence, which affects up to 40% of kidney transplant recipients. The development of novel DSA post-transplant should trigger an allograft biopsy to exclude rejection.

Overall, the rate of de novo DSA development is 15–20% at 5 years post-transplant, and it is associated with worse graft survival, affecting especially younger patients, non-compliant, and/or with greater DR mismatches [17]. The mechanism of injury is through damage to the endothelium of vascular beds, including peritubular capillaries, small-medium sized vessels, and glomeruli. Continuous antibody-mediated injury leads to basement membrane duplication (transplant glomerulopathy on graft biopsy), podocyte injury, and cellular hypertrophy, indicating chronicity. In the absence of inflammation such as peritubular capillaritis or glomerulitis, these lesions are non-specific to rejection as there are other causes such as recurrent autoimmune glomerulonephritis, thrombotic microangiopathy, and hepatitis C-related glomerulonephritis that may lead to similar endothelial injury pattern.

One of the key components of the DeKAF study was to evaluate for-cause biopsies in kidney transplant recipients on average 7 years post-transplant. The authors found that 99 of 173 patients (57%) with allograft dysfunction were due to antibody-mediated injury and the severity of injury correlated with the intensity of the antibody response [18]. Several techniques have been developed to improve the detection of ABMR, including single HLA antigen bead assay, C4d staining, electron microscopy to assess endothelial injury, and gene profiling in allograft biopsy samples. Challenging cases include those with no circulating DSA but clear picture of ABMR or those with C4d-negative ABMR.

Creatinine is an insensitive biomarker for kidney injury and challenging to interpret in the setting of kidney transplantation. Non-invasive biomarkers can assist in identifying early allograft injury and predicting treatment response and earlier interventions. For rejection, protein and mRNA levels of CXCL9 can be detected even 30 days prior to biopsy-proven rejection [19]. Given its high negative predictive value (92%), it could be used for risk stratification for allograft biopsy. A 3 gene signature (urinary cell mRNA) of CD3E, IP-10, and 18S ribosomal RNA is diagnostic of acute rejection and capable to distinguish between cellular and antibody-mediated rejection [20]. The only biomarker of rejection in clinical practice today is the measurement of anti-HLA DSA that may help in identifying patients at risk for antibody-mediated rejection, although the methods used to measure it have to be interpreted with caution as not all patients with DSA will develop ABMR.

Glomerular Disease Recurrence

In the Australia/New Zealand registry [21], recurrence was the third leading cause of end-stage renal disease in kidney transplant recipients, with a 10 year incidence of 8.4%. The type of immunosuppression does not seem to be a risk factor for recurrence [22].

Focal segmental glomerulosclerosis (FSGS) may present early after transplant with massive proteinuria leading to graft loss in over 50% of the cases. Etiology is unclear; however, the presence of a circulating factor leading to podocyte injury is suspected [23]. Recent study suggested that genetic analyses may be a reasonable approach to help stratify the risk of recurrence as evident by the lack of recurrence in all patients with genetic-identified primary FSGS [24].

Another recurrent disease (30–90%) with poor prognosis is atypical hemolytic-uremic syndrome; severity depends on the underlying mutation, for instance, the risk is low (~15%) for mutations in membrane cofactor protein and high (~80%) for mutations in circulating proteins [25]. IgA nephropathy recurs in up to 40% of the cases but rarely causes early graft loss. However, risk of clinically significant recurrence increases over time [26, 27]. Recurrent membranous nephropathy is typically observed 13–15 months after transplantation, although it may be observed much earlier (within weeks up to 2 years) [28]. Lastly, C3 glomerulonephritis also tends to recur. Zand et al. [29] reported that the median time of disease recurrence was 2.3 years and 50% of the patients lost their grafts in an average 6.4 years after kidney transplantation. It usually presents with graft dysfunction, hematuria, proteinuria, and hypocomplementemia, and once it recurs, often results in graft loss. In summary, knowing the underlying etiology of kidney disease may help determine the risk of recurrence though further investigation will be required to improve the precision of this prediction, probably with the development of biomarkers and performance of genetic analysis.

Infections (Viral/Bacterial)

With the increased use of immunosuppressive drugs during the last 3 decades, BK nephropathy has become a major cause of acute and chronic allograft dysfunctions. BK is an ubiquitous virus, tropic to the renal epithelium that proliferates in the setting of immunosuppression. Immunosuppression intensity is the main risk factor and donor and recipient BK serologies do not predict infection risk. BKV first causes viruria, followed by viremia and nephropathy and if kidney recipients are affected during the first year post-transplant with BK nephropathy, 30–65% of the cases may lose their grafts [30] Therefore, surveillance of BKV PCR is imperative as early BKV detection may warrant reduction of immunosuppression. Greater than 10,000 copies of BKV with allograft dysfunction, is likely due to BK nephropathy. A biopsy with SV40 staining confirms the diagnosis.

Another common infection in the post-transplant period is cytomegalovirus (CMV) that may present as CMV syndrome (fever, weakness, myalgia) or end-organ disease (pneumonitis, hepatitis, colitis). CMV infection involving the kidney allograft is rare, but it could cause necrotizing crescentic glomerulonephritis with viral inclusions and CMV-induced hemophagocytic syndrome. Lastly, bacterial infections such as UTIs, which are frequently caused by gram-negative organisms, may lead to chronic damage due to direct injury to the allograft or indirectly by activation of the innate immunity, increasing the expression of co-stimulatory molecules and lowering the threshold for allo-T cell activation and proliferation against the allograft.

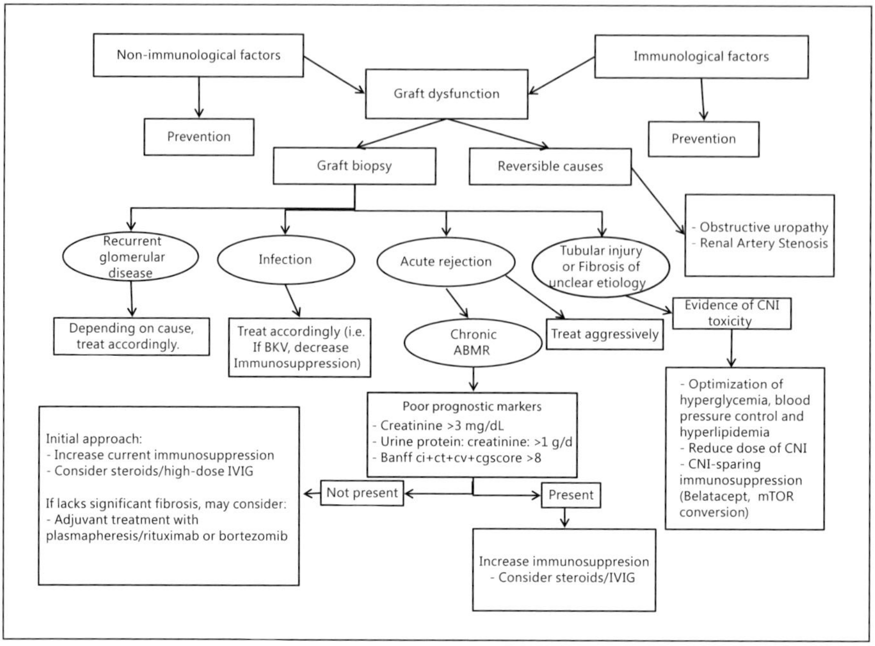

Fig. 1. Proposed algorithm for chronic graft dysfunction management after kidney transplantation.

Diagnosis of CAI

Obtaining a detailed history of the primary renal disease, early post-transplant course, episodes of acute rejection, hypertension/glycemic control, overview of CNI levels, inquiring about adherence, and microscopic urinalysis may support in the identification of potential etiologies of CAI (Figure 1). Cellular casts, granular casts or significant proteinuria, may indicate an active glomerulonephritis, acute tubular necrosis or transplant glomerulopathy, respectively. An allograft ultrasound excludes obstructive causes such as ureteral stenosis and vascular abnormalities. A biopsy should be performed anytime there is an acute graft dysfunction, proteinuria, cellular casts, and no obvious cause of allograft dysfunction. Concomitant testing for DSA should be performed to help identify ABMR.

The role of protocol biopsies is debated. Its goal is to detect unsuspected graft injury, which may lead to change in management in up to 25% of the cases [31]. Subclinical antibody-mediated rejection is associated with worse 8-year outcomes (56%) compared to subclinical cellular-mediated rejection (88%). Early post-transplant subclinical inflammation may be associated with worse graft survival, and it is important to consider that not all infiltrating cells (such as Tregs) present in subclinical inflammation are pathogenic. Treating all cases of subclinical or borderline rejection with increased immunosuppression is unwarranted at this time as there is no clear evidence that this approach is associated with improved graft survival.

Prevention of CAI

Blood pressure control post-transplant is linked to improved graft survival at 1 and 3 years post-transplant, and each 20-mm Hg increase in systolic blood pressure is associated with increased risk of cardiovascular disease and death [32]. ACEi/ARBs are renoprotective in the general population with kidney disease, though a randomized controlled trial in transplantation did not show any benefit [33]. Other anti-hypertensive agents such as amlodipine are routinely recommended as first-line agents but the ideal goal of blood pressure remains unknown.

Belatacept, is an alternative to CNI that has shown promise in recent clinical trials. Belatacept is a CTLA4Ig fusion protein that blocks the CD80/CD86 interaction with CD28, preventing T cell activation. This maintenance medication is contraindicated in patients with negative EBV status for concerns of a higher risk of post-transplant lymphoproliferative disorder. Albeit the FDA approval of belatacept for kidney transplantation in 2011, its association with high rate of acute cellular rejection in the first year post-transplant has limited its use among transplant centers. Despite this, the 7-year follow-up data with belatacept revealed better graft function and patient survival when compared to cyclosporine [34]. One of the limitations of the trial is the comparison to cyclosporine, instead of tacrolimus. Furthermore, only a small portion of patients initially randomized to Belatacept completed the 7-year follow-up, creating a selection bias. Finally, novel strategies to mitigate the higher acute rejection with belatacept are under investigation, including the addition of CNIs temporarily during first year after transplant, the use of depletion induction therapy, and the conversion of mycophenolic acid to mTOR inhibitors.

With regard to mTOR inhibitors, trials have not shown any benefit in improving long-term graft survival when substituting CNI. In reality, conversion from CNI to mTOR inhibitor was associated with greater risk of rejection and DSA development [35, 36].

To minimize injury from cold ischemia, machine perfusion has been utilized to reduce the risk of delayed graft function and improve graft survival in the first year after transplantation [37].

Lastly, non-adherence prevention is another element that needs to be addressed. Several risk factors are involved such as young age, fluctuating drug levels, missed appointments, stress, financial crisis and psychosocial factors such as poor social support, and psychiatric disorders. Screening at every clinic visit, especially in high risk individuals, and understanding each patient's individual needs, is crucial for medication adherence. Acknowledging the challenges and burden of taking immunosuppressive drugs is important so that strategies to minimize errors are developed.

Therapeutic Options of CAI

Treating chronic ABMR is difficult and the best approach would be to avoid it by adequately treating acute ABMR. Therapeutic options for acute ABMR include plasmapheresis with intravenous immunoglobulin (IVIG). Adjuvant agents that may be used include: immunoadsorption, bortezomib, rituximab, or eculizumab. However even after standard therapy with IVIG/plasmapheresis with rituximab, more than 40% of the cases are not resolved [38].

With regard to chronic ABMR, treatment options are limited and prognosis after diagnosis tends to be poor. Redfield et al. [39] performed a retrospective study on 123 patients, using the Banff 2013 criteria, and found that 76% of grafts were lost with a median survival of 1.9 years after diagnosis of chronic ABMR. Treatment of chronic ABMR has been derived from acute ABMR treatment, though it seems less effective. A combination of Rituximab/IVIG improved the graft function in a small group of patients [40]. In contrast, other studies have used plasma exchange, IVIG, and rituximab without good outcomes [41]. In a study of 31 kidney transplant recipients with chronic ABMR, median survival of patients treated with rituximab therapy experienced longer allograft survival (22.5 months) in comparison to controls (14.4 months) [42].

Steroids/IVIG is another therapeutic option with mixed results, and adding rituximab or thymoglobulin® to the regimen has shown no improvement in outcomes or graft survival [39]. In conclusion, there are no good available therapeutic options for the treatment of ABMR and interventional clinical trials addressing safety and efficacy of different DSA and B cell depleting agents are needed.

Recurrent Primary Glomerulonephritis

Recurrent IgA nephropathy may be treated with ACEI/ARB therapy, and addition of oral prednisone (with taper) to the regimen may be advised [43].

There is limited evidence for the treatment of recurrent membranous nephropathy; however, some patients may respond to rituximab [44, 45]. In ritux-

Table 2. Treatment of recurrent glomerulonephritis in kidney transplantation

Recurrent disease in graft	Proposed therapeutic approaches (most without solid evidence)
Focal segmental glomerulosclerosis	Steroids, cyclosporine (high dose), plasma exchange
IgA nephropathy	RAS inhibitors, steroids
Membranous nephropathy	Rituximab
Membranoproliferative	Steroids, cyclophosphamide, plasma exchange, rituximab, eculizumab
Hemolytic uremic syndrome (recurrent or de novo)	Eculizumab, plasmapheresis
C3 glomerulopathy (DDD)	Steroids, rituximab, plasmapheresis, eculizumab

imab-resistant cases, cyclophosphamide may be attempted, although there is no conclusive evidence.

For FSGS recurrence, therapeutic strategies include high dose steroids and cyclosporine (for podocyte stabilization) in combination with plasma exchange (to remove circulating humoral factors) [46, 47].

Treatment of recurrent membranoproliferative glomerulonephritis (MPGN) is directed at the underlying etiology. Anecdotal cases of type I or idiopathic have been treated successfully with rituximab [48–50], although other therapeutic options include steroids, cyclophosphamide and/or plasma exchange. Eculizumab has been used in cases where MPGN is caused by dysregulation of the alternative complement pathway, including C3 glomerulonephritis [29] (Table 2).

Conclusion

CAI is a major cause of graft loss and its prevention should be the focus of care of kidney transplant recipients. New tools are currently being developed such as a gene expression analysis that could identify patients at risk, so an early prophylactic plan can be enforced. Some of the strategies that we can implement now include: aggressive treatment of modifiable risk factors such as hypertension, hyperglycemia, hyperlipidemia, CNI minimization, protocol biopsies, addressing drug adherence, and performing surveillance BKV and DSA measurements post-transplant.

References

1 Solez K, Colvin RB, Racusen LC, Sis B, Halloran PF, Birk PE, et al: Banff '05 Meeting Report: differential diagnosis of chronic allograft injury and elimination of chronic allograft nephropathy ("CAN"). Am J Transplant 2007;7:518–526.

2 Lamb KE, Lodhi S, Meier-Kriesche HU: Long-term renal allograft survival in the United States: a critical reappraisal. Am J Transplant 2011;11:450–462.

3 El-Zoghby Z, Stegall MD, Lager DJ, Kremers WK, Amer H, Gloor JM, et al: Identifying specific causes of kidney allograft loss. Am J Transplant 2009;9:527–535.

4 Ekberg H, Tedesco-Silva H, Demirbas A, Vítko S, Nashan B, Gürkan A, et al: Reduced exposure to calcineurin inhibitors in renal transplantation. N Engl J Med 2007;357:2562–2575.

5 Ekberg H, Bernasconi C, Tedesco-Silva H, Vítko S, Hugo C, Demirbas A, et al: Calcineurin inhibitor minimization in the Symphony study: observational results 3 years after transplantation. Am J Transplant 2009;9:1876–1885.

6 Snanoudj R, Royal V, Elie C, Rabant M, Girardin C, Morelon E, et al: Specificity of histological markers of long-term CNI nephrotoxicity in kidney-transplant recipients under low-dose cyclosporine therapy. Am J Transplant 2011;11:2635–2646.

7 Hricik DE, Formica RN, Nickerson P, Rush D, Fairchild RL, Poggio ED, et al: Adverse outcomes of tacrolimus withdrawal in immune-quiescent kidney transplant recipients. J Am Soc Nephrol 2015;26:3114–3122.

8 Freedman BI, Pastan SO, Israni AK, Schladt D, Julian BA, Gautreaux MD, et al: APOL1 genotype and kidney transplantation outcomes from deceased African American donors. Transplantation 2016;100:194–202.

9 Riella LV, Sheridan AM: Testing for high-risk APOL1 alleles in potential living kidney donors. Am J Kidney Dis 2015;66:396–401.

10 Beckerman P, Bi-Karchin J, Park AS, Qiu C, Dummer PD, Soomro I, et al: Transgenic expression of human APOL1 risk variants in podocytes induces kidney disease in mice. Nat Med 2017, Epub ahead of print.

11 Giral M, Foucher Y, Karam G, Labrune Y, Kessler M, Hurault de Ligny B, et al: Kidney and recipient weight incompatibility reduces long-term graft survival. J Am Soc Nephrol 2010;21:1022–1029.

12 Yokoyama I, Uchida K, Kobayashi T, Tominaga Y, Orihara A, Takagi H: Effect of prolonged delayed graft function on long-term graft outcome in cadaveric kidney transplantation. Clin Transplant 1994;8(2 pt 1):101–106.

13 Weir MR, Burgess ED, Cooper JE, Fenves AZ, Goldsmith D, McKay D, et al: Assessment and management of hypertension in transplant patients. J Am Soc Nephrol 2015;26:1248–1260.

14 Guijarro C, Massy ZA, Kasiske BL: Clinical correlation between renal allograft failure and hyperlipidemia. Kidney Int Suppl 1995;52:S56–S59.

15 Bagley J, Yuan J, Chandrakar A, Iacomini J: Hyperlipidemia alters regulatory T cell function and promotes resistance to tolerance induction through costimulatory molecule blockade. Am J Transplant 2015;15:2324–2335.

16 Yuan J, Bagley J, Iacomini J: Hyperlipidemia promotes anti-donor Th17 responses that accelerate allograft rejection. Am J Transplant 2015;15:2336–2345.

17 Wiebe C, Gibson IW, Blydt-Hansen TD, Karpinski M, Ho J, Storsley LJ, et al: Evolution and clinical pathologic correlations of de novo donor-specific HLA antibody post kidney transplant. Am J Transplant 2012;12:1157–1167.

18 Gaston RS, Cecka JM, Kasiske BL, Fieberg AM, Leduc R, Cosio FC, et al: Evidence for antibody-mediated injury as a major determinant of late kidney allograft failure. Transplantation 2010;90:68–74.

19 Hricik DE, Nickerson P, Formica RN, Poggio ED, Rush D, Newell KA, et al: Multicenter validation of urinary CXCL9 as a risk-stratifying biomarker for kidney transplant injury. Am J Transplant 2013;13:2634–2644.

20 Suthanthiran M, Muthukumar T: Urinary-cell mRNA and acute kidney-transplant rejection. N Engl J Med 2013;369:1860–1861.

21 Briganti EM, Russ GR, McNeil JJ, Atkins RC, Chadban SJ: Risk of renal allograft loss from recurrent glomerulonephritis. N Engl J Med 2002;347:103–109.
22 Mulay AV, van Walraven C, Knoll GA: Impact of immunosuppressive medication on the risk of renal allograft failure due to recurrent glomerulonephritis. Am J Transplant 2009;9:804–811.
23 Gallon L, Leventhal J, Skaro A, Kanwar Y, Alvarado A: Resolution of recurrent focal segmental glomerulosclerosis after retransplantation. N Engl J Med 2012;366:1648–1649.
24 Bierzynska A, McCarthy HJ, Soderquest K, Sen ES, Colby E, Ding WY, et al: Genomic and clinical profiling of a national nephrotic syndrome cohort advocates a precision medicine approach to disease management. Kidney Int 2017;pii:S0085-2538(16)30612-3.
25 Le Quintrec M, Lionet A, Kamar N, Karras A, Barbier S, Buchler M, et al: Complement mutation-associated de novo thrombotic microangiopathy following kidney transplantation. Am J Transplant 2008;8:1694–1701.
26 Moroni G, Longhi S, Quaglini S, Gallelli B, Banfi G, Montagnino G, et al: The long-term outcome of renal transplantation of IgA nephropathy and the impact of recurrence on graft survival. Nephrol Dial Transplant 2013;28:1305–1314.
27 Odum J, Peh CA, Clarkson AR, Bannister KM, Seymour AE, Gillis D, et al: Recurrent mesangial IgA nephritis following renal transplantation. Nephrol Dial Transplant 1994;9:309–312.
28 Dabade TS, Grande JP, Norby SM, Fervenza FC, Cosio FG: Recurrent idiopathic membranous nephropathy after kidney transplantation: a surveillance biopsy study. Am J Transplant 2008;8:1318–1322.
29 Zand L, Lorenz EC, Cosio FG, Fervenza FC, Nasr SH, Gandhi MJ, et al: Clinical findings, pathology, and outcomes of C3GN after kidney transplantation. J Am Soc Nephrol 2014;25:1110–1117.
30 Randhawa P, Brennan DC: BK virus infection in transplant recipients: an overview and update. Am J Transplant 2006;6:2000–2005.
31 Henderson LK, Nankivell BJ, Chapman JR: Surveillance protocol kidney transplant biopsies: their evolving role in clinical practice. Am J Transplant. 2011;11:1570–1575.
32 Carpenter MA, John A, Weir MR, Smith SR, Hunsicker L, Kasiske BL, et al: BP, cardiovascular disease, and death in the folic acid for vascular outcome reduction in transplantation trial. J Am Soc Nephrol 2014;25:1554–1562.
33 Knoll GA, Fergusson D, Chassé M, Hebert P, Wells G, Tibbles LA, et al: Ramipril versus placebo in kidney transplant patients with proteinuria: a multicentre, double-blind, randomised controlled trial. Lancet Diabetes Endocrinol 2016;4:318–326.
34 Vincenti F, Rostaing L, Grinyo J, Rice K, Steinberg S, Gaite L, et al: Belatacept and Long-term outcomes in kidney transplantation. N Engl J Med 2016;374:333–343.
35 Croze LE, Tetaz R, Roustit M, Malvezzi P, Janbon B, Jouve T, et al: Conversion to mammalian target of rapamycin inhibitors increases risk of de novo donor-specific antibodies. Transpl Int 2014;27:775–783.
36 Schena FP, Pascoe MD, Alberu J, del Carmen Rial M, Oberbauer R, Brennan DC, et al: Conversion from calcineurin inhibitors to sirolimus maintenance therapy in renal allograft recipients: 24-month efficacy and safety results from the CONVERT trial. Transplantation 2009;87:233–242.
37 Moers C, Smits JM, Maathuis MH, Treckmann J, van Gelder F, Napieralski BP, et al: Machine perfusion or cold storage in deceased-donor kidney transplantation. N Engl J Med 2009;360:7–19.
38 Sautenet B, Blancho G, Büchler M, Morelon E, Toupance O, Barrou B, et al: One-year results of the effects of rituximab on acute antibody-mediated rejection in renal transplantation: RITUX ERAH, a multicenter double-blind rRandomized Placebo-controlled Trial. Transplantation 2016;100:391–399.
39 Redfield RR, Ellis TM, Zhong W, Scalea JR, Zens TJ, Mandelbrot D, et al: Current outcomes of chronic active antibody mediated rejection – a large single center retrospective review using the updated BANFF 2013 criteria. Hum Immunol 2016,77:346–352.
40 Fehr T, Rüsi B, Fischer A, Hopfer H, Wüthrich RP, Gaspert A: Rituximab and intravenous immunoglobulin treatment of chronic antibody-mediated kidney allograft rejection. Transplantation 2009;87:1837–1841.

41 Waiser J, Budde K, Schütz M, Liefeldt L, Rudolph B, Schönemann C, et al: Comparison between bortezomib and rituximab in the treatment of antibody-mediated renal allograft rejection. Nephrol Dial Transplant 2012;27:1246–1251.
42 Smith RN, Malik F, Goes N, Farris AB, Zorn E, Saidman S, et al: Partial therapeutic response to Rituximab for the treatment of chronic alloantibody mediated rejection of kidney allografts. Transpl Immunol 2012;27: 107–113.
43 Lv J, Xu D, Perkovic V, Ma X, Johnson DW, Woodward M, et al: Corticosteroid therapy in IgA nephropathy. J Am Soc Nephrol 2012; 23:1108–1116.
44 Fervenza FC, Cosio FG, Erickson SB, Specks U, Herzenberg AM, Dillon JJ, et al: Rituximab treatment of idiopathic membranous nephropathy. Kidney Int 2008;73:117–125.
45 Sprangers B, Lefkowitz GI, Cohen SD, Stokes MB, Valeri A, Appel GB, et al: Beneficial effect of rituximab in the treatment of recurrent idiopathic membranous nephropathy after kidney transplantation. Clin J Am Soc Nephrol 2010;5:790–797.
46 Canaud G, Martinez F, Noël LH, Mamzer MF, Niaudet P, Legendre C: Therapeutic approach to focal and segmental glomerulosclerosis recurrence in kidney transplant recipients. Transplant Rev (Orlando) 2010;24: 121–128.
47 Staeck O, Slowinski T, Lieker I, Wu K, Rudolph B, Schmidt D, et al: Recurrent primary focal segmental glomerulosclerosis managed with intensified plasma exchange and concomitant monitoring of soluble urokinase-type plasminogen activator receptor-mediated podocyte beta3-integrin activation. Transplantation 2015;99:2593–2597.
48 Farooqui M, Alsaad K, Aloudah N, Alhamdan H: Treatment-resistant recurrent membranoproliferative glomerulonephritis in renal allograft responding to rituximab: case report. Transplant Proc 2015;47:823–826.
49 Guiard E, Karras A, Plaisier E, Duong Van Huyen JP, Fakhouri F, Rougier JP, et al: Patterns of noncryoglobulinemic glomerulonephritis with monoclonal Ig deposits: correlation with IgG subclass and response to rituximab. Clin J Am Soc Nephrol 2011;6: 1609–1616.
50 Lorenz EC, Sethi S, Leung N, Dispenzieri A, Fervenza FC, Cosio FG: Recurrent membranoproliferative glomerulonephritis after kidney transplantation. Kidney Int 2010;77:721–728.

Leonardo V. Riella, MD, PhD
Renal Division, Brigham and Women's Hospital
Harvard Medical School
221 Longwood Ave., Boston, MA 02115 (USA)
E-Mail lriella@bwh.harvard.edu

La Manna G, Ronco C (eds): Current Perspectives in Kidney Diseases.
Contrib Nephrol. Basel, Karger, 2017, vol 190, pp 181–193 (DOI: 10.1159/000468967)

Looking into the Graft without a Biopsy: Biomarkers of Acute Rejection in Renal Transplantation

Andrea Angeletti[a, b] • Paolo Cravedi[a]

[a]Department of Medicine, Translational Transplant Research Center, Recanati Miller Transplant Institute, Icahn School of Medicine at Mount Sinai, New York, NY, USA; [b]Department of Experimental Diagnostic and Specialty Medicine (DIMES), Nephrology, Dialysis and Renal Transplant Unit, St. Orsola Hospital, University of Bologna, Bologna, Italy

Abstract

Immunosuppression in kidney transplantation is still largely driven by center-specific protocols and adjusted according to the presence of graft dysfunction, drug toxicity or infection. Due to the current inability to titrate immunosuppression to the needs of every single patient, this approach is burdened by an increased risk of under or over-immunosuppression. To extend allograft survival, immunosuppression should be decided based on reliable biomarkers capable of quantifying the alloimmune response and/or noninvasively diagnosing acute rejection. Emerging data indicate that many assays, likely used in panels rather than singularly, have potential to be diagnostic and predictive of short- and also long-term outcomes. Many cross-sectional and retrospective studies showed associations between these biomarkers and clinically relevant post-transplantation outcomes, but data from prospective studies are still scarce, thereby preventing widespread implementation in the clinic. Prospective, randomized, multicenter studies are currently ongoing to test the hypothesis that biomarker-driven immunosuppression provides better outcomes than standard, protocol-driven immunosuppression. These projects and other future studies are highly warranted to improve outcomes of kidney transplant patients.

Introduction

Despite advancements in transplant care and excellent short-term outcomes, long-term allograft survival rates remain suboptimal [1]. Management of transplant patients is generally initiated by protocols and adjusted based on graft function, immunosuppressive drug levels in blood or patient infection. However, the recognition of clinical, immunological, and genetic heterogeneity that exists among donors and recipients raises the possibility that immunosuppression could be tailored to optimize the outcome for each recipient. Currently, immunosuppression is guided by relatively rudimentary strategies including epidemiologic parameters (e.g., age and self-declared race), human leukocyte antigen (HLA) mismatch, and donor-specific anti-HLA antibody (DSAs) screening. Development of individualized treatment strategies requires reliable, reproducible, cost effective, yet noninvasive biomarkers capable of quantifying the alloimmune response and predicting and/or diagnosing allograft injury. Ideally, these biomarkers guide immunosuppressant withdrawal in patients at minimal risk and direct specific therapeutic strategies to limit injury in patients at higher risk, together resulting in prolonged graft survival and improved patient health. The purpose of this review is to summarize recent progress in biomarker research relevant to kidney transplantation.

Biomarker Fundamentals

In kidney transplantation, biomarkers could be used for assessing organ quality pre-transplant, predicting/diagnosing acute rejection (AR), identifying immunologic/operational tolerance, over-immunosuppression, nonadherence, and opportunistic infections, or estimating the risk of progressive, chronic graft injury [2].

An ideal immune biomarker should be based on a simple and inexpensive assay, requiring a noninvasively obtained sample (e.g., urine or blood), yet it should be specific, sensitive, and able to predict a clinically relevant endpoint. Biomarker discovery can be obtained via "biased" and "unbiased" approaches. Biased discovery starts from a basic science mechanism and tests whether the involved molecules or readouts can predict a clinical endpoint. It has the advantage of being supported by previous scientific findings but can be limited in scope [3]. Unbiased approaches attempt to correlate molecular patterns derived from genomic, proteomic, or other large-scale screening assays with clinical phenotypes. While this strategy is hypothesis-generating rather than hypothesis-testing, it can result in the discovery of novel biomarkers that can drive new mechanistic studies [3]. For any biomarker to become a useful test in the clinic, (1) it has to be first prospectively validated in independent large patient cohorts, (2) any qualified laboratory should be able to replicate the results, and (3) it should be commercially available for routine clinical use.

Quantifying Alloreactive Antibodies

The effector role of anti-HLA antibodies in graft loss has been known since 1960s, when it was observed that sensitized transplant patients develop hyper AR [4]. Even though the issue of hyperacute allograft rejection has been overcome by the introduction of novel crossmatch techniques, humoral immunity against donor antigens still represents a major cause of late allograft loss: once produced, anti-HLA antibodies bind to donor tissue and mediate damage through multiple mechanisms including, among others, complement activation [5].

Numerous studies indicated that solid-phase assays to measure anti-HLA antibodies have the highest sensitivity for predicting post-transplant antibody-mediated rejection (ABMR) and/or graft loss [6–8]. Standardization of solid-phase assay techniques among laboratories is achievable but the assay is not designed to be quantitative. Though transplant centers routinely avoid transplantation with high titer of donor-specific antibodies (mean fluorescence intensity [MFI] of >10,000), the thresholds for defining positivity and the clinical implications of pre-transplant DSAs with lower MFI remain controversial [9]. Evidence from multiple studies associates development of *de novo* post-transplant DSAs with an elevated risk of late graft loss [8], particularly in nonadherent patients [10]. Reports also suggest a higher risk for kidney allograft loss in subjects with serum DSA that bind/activate complement as measured by a standard solid-phase testing assay that additionally detects C1q binding [6–8]. In a prospective study in 1,016 kidney transplant patients [6], C1q-positive de novo DSA were associated with a shorter graft survival than C1q-negative de novo DSA or the absence of any DSA [10]. It is still unclear whether C1q positivity indicates antibodies preferentially capable of initiating complement-dependent allograft rejection, or if C1q positivity is just the result of higher DSA titers [7] rather than complement-activating activity of the DSA per se. Therefore, the clinical utility of C1q binding DSAs as a risk assessment tool and the timing/frequency of DSA testing for detection of de novo DSAs remains uncertain.

C1q test is an expression of antibody binding to a purified C1q molecule and does not necessarily imply activation of the complement cascade. The C3d assay, based on the addition of human serum as a source of complement, constitutes a better measurement of complement cascade activation. In a prospective cohort study on 114 pediatric not sensitized first kidney allografts [11], we analyzed post-transplant kinetics of complement-fixing de novo DSAs and their role in ABMR development and graft loss. Overall, 39 patients developed de novo DSAs, which were activating C1q and C3d in 25 and 9 patients, respectively. The 10-year graft survival probability was lower in patients with C3d-binding DSA than in those without DSA, with C1q but not C3d activating or with non-complement-binding DSA. This study indicates that it is reasonable to intensify treatment at first confirmed positivity for C3d in de novo DSA.

Quantifying Memory B Cells

Measuring circulating antibodies to determine the allosensitization state of a given transplant patient may underestimate the magnitude of the complete humoral immune response. The absence of alloantibodies does not exclude the presence of alloreactive B cells. Two major B-cell populations contribute to the maintenance of immunological memory: long-lived plasma cells and memory B cells (mBCs). Alloreactive mBCs may exist in the absence of detectable serum antibody levels, and their rapid differentiation and antibody production are important for a protective antibody response against pathogens [12]. Therefore, the detection of functionally active anti-donor HLA-mBCs could provide additional relevant biological information regarding the sensitization state of transplant candidates. Different approaches have been developed to measure alloreactive mBCs (Table 1).

Flow Cytometry

Taking advantage of the introduction of the HLA tetramer technology to quantify antigen-specific T cells, Mulder et al. [13], using both human engineered B cells and HLA-specific B cells from pregnancy-immunized individuals, were the first to show the feasibility of tracking HLA-specific B cells binding epitopes of foreign HLA molecules in their original conformation through their B-cell receptor. This technology uses streptavidin-biotin complexes of HLA molecules conjugated to a fluorescent protein. Then, the tetramer-binding B cells can be accurately enumerated by flow cytometry, allowing a rapid quantization of the B-cell response to the given HLA antigen. To discriminate the tetramer-binding mBC and plasma cells, Zachary et al. [14], added CD27

Table 1. Most promising immune monitoring assays

Assay	Description	Pros/Cons
Alloreactive B cells		
HLA-specific tetramer-binding B cells	HLA-specific B cells are detected by fluorochrome-conjugated HLA tetramers using flow cytometry; the HLA tetramers are streptavidin-biotin complexes of 4 peptide-loaded HLA molecules conjugated to a fluorescent protein	Pros: relatively easy detection of B cells harboring a HLA-sp B-cell receptor Cons: low sensitivity to detect low frequencies of alloreactive B cells; does not identify mBC capable of secreting alloantibodies
Anti-HLA antibodies from ex-vivo expanded mBC culture supernatants	Indirect assessment of HLA-sp mBC by measuring anti-HLA antibodies in supernatants of ex-vivo expanded mBC; PBMC containing mBC are activated for 6 days and alloantibodies in the supernatants are detected by Luminex	Pros: simple detection of anti-HLA Ab released by mBC Cons: indirect quantification; Low sensitivity for low Ab titers (effect of dilution); needs 6-day polyclonal stimulation
HLA-specific B-cell ELISPOT assay	Quantifies the frequency of anti-HLA mBC; after 6-day polyclonal activation, cells are seeded in ELISPOT wells pre-coated with multimerized biothynilated HLA class I or II molecules; spots that develop (representing mBC producing anti-HLA antibodies) are quantified using computerized plate readers	Pros: precise quantification of alloreactive mBCs Cons: 6-day cell culture to induce and differentiate circulating mBCs; the need of the whole HLA antigen repertoire to be tested in these assays; expensive and labor-intensive
Alloreactive T cells		
Enzyme-linked immunospot (ELISPOT)	Recipient PBMC are cultured for 18–24 h in the presence of inactivated donor or third-party cells in tissue culture wells coated with a capture antibody against the cytokine of interest; spots that develop represent cells that had been primed to the stimulating antigen(s) in vivo	Pros: high throughput and functional readout; faster than MLR and amenable to standardization; multiple prospective studies support its predictive value Cons: expensive and labor-intensive
Panel of reactive T-cells (PRT)	The assay quantifies peripheral effector-memory T cells reacting against a pool of 6–8 B cell donors thought to be representative of the pool of potential donors	Pros: does not require donor cells (potentially applicable for deceased donor recipients) Cons: not yet investigated in large cohorts of kidney transplant patients
Urinary biomarkers		
C-X-C motif chemokine ligand 9 (CXCL9)	Recipient urine supernatants are tested by ELISA for CXCL9	Pros: easily measurable, reliable marker to exclude AR and to stratify patients between low versus high-risk for allograft injury Cons: urinary tract infections can elevate urinary CXCL9
mRNA	Gene expression in urinary cells quantified by real-time polymerase chain reaction (qRT-PCR)	Pros: high specificity Cons: labor-intensive assay requiring well-trained technicians, lack of inter-laboratory standard values
Molecular biomarkers		
Kidney solid organ response test (kSORT)	17-gene qPCR-based assay in recipient blood samples	Pros: relatively easy to perform and standardize Cons: lack of multicenter validation
Cell-free DNA		
Graft-derived cell-free DNA (GcfDNA)	Droplet digital PCR can be used to quantify GcfDNA in the plasma of organ transplant recipients	Pros: reasonable costs, requires limited very low amounts of plasma making it suitable for pediatric patients Cons: cannot distinguish AR from other causes of graft injury

mBC, memory B cells; HLA-sp, human leukocyte antigen-specific; Ab, antibody; Ig, immunoglobulin; ELISA, Enzyme-Linked Immunosorbent Assay.

and CD38 markers, respectively. In 16 renal transplant candidates with no DSA at the time of transplantation, the presence of anti-donor HLA-tetramer positive mBCs was associated with the subsequent development of DSAs after transplantation. This association was more evident in patients who did not receive anti-CD20 B-cell depleting antibody rituximab [15]. A main limitation of such technique is that >6% B cells can recognize non-HLA fractions of the tetramer, such as the streptavidin-phycoerythrin, leading to unspecific staining [16].

Measuring Anti-HLA Antibodies in Supernatants of ex vivo Expanded mBC Cultures

Ex vivo expansion of mBCs and measurement of anti-HLA antibodies in culture supernatants represent an alternative approach to indirectly quantify the presence of alloreactive mBCs. By using this strategy, Han et al. [17] were able to identify mBC producing HLA-specific antibodies in 12 sensitized organs transplant recipients, but not in not-sensitized controls. Importantly, DSAs were found in some B-cell cultures but not in concomitant serum of patients, confirming such DSAs were released by circulating mBCs rather than by long-living plasma cells. In a recent report, Snanoudj et al. [18] found the presence of HLA-specific antibodies in expanded mBC culture supernatants in 46% of 39 highly sensitized kidney transplant recipients, showing a high burden of previous allogeneic sensitization, even long time after the first allogeneic contact. Such results are highly suggestive of how the expansion/selection/survival of oligoclonal alloreactive mBCs would be impacted by dose and duration of exposure to alloantigens.

Enzyme-Linked Immunosorbent Spot Assays

To more accurately quantify the frequency of HLA-specific mBCs, Bestard's group developed a B-cell enzyme-linked immunosorbent spot (ELISPOT) assay platform [19]. This assay is capable of precisely quantifying the numbers of alloantibody secreting cells. In a study including 70 kidney transplant recipients, Lúcia et al. [20] found circulating mBCs against both class I and class II HLA antigens in patients with allogeneic sensitization background, even in the absence of detectable circulating anti-HLA antibodies. In the same study, donor-specific alloreactive mBCs were present in all 16 kidney transplant patients during ABMR and also before transplantation, reflecting an active baseline anti-donor sensitization state not always depicted by circulating DSA. Currently, 2 large European observational studies (Leiden, Holland and Barcelona, Spain) are ongoing to assess the value of quantifying HLA-specific mBC in kidney transplantation [21]. However, the 6-day cell culture and the use of cells as the biological specimen to analyze are major hurdles that may limit its implementation in the clinical practice. Furthermore, the need for the whole HLA antigen repertoire to be tested in these assays is required to cover all potential donor-recipient HLA mismatches of all transplant candidates around the globe.

Measuring Alloreactive Memory T Cells

Preformed alloreactive memory T cells represent a major risk factor for AR and accelerated graft loss [22]. Therefore, measurements of memory alloimmunity could be used as a risk stratifier for AR after transplant (Table 1).

Alloreactive T-Cell ELISPOT Assay

This assay provides a measure of the frequency of previously activated or memory T cells that respond to specific antigens by producing a selected cytokine (for more details, refer to [23]). Pre-transplantation measurements of recipient T-cell alloreactivity to donor antigen via the IFN-γ ELISPOT assay are consistently correlated with AR after kidney transplantation [23–25]. ELISPOT also has potential utility in the identification of patients who may benefit the most from antibody induction therapy. In a retrospective analysis of 130 kidney transplant patients, individuals with higher pre-transplantation ELISPOT levels were at increased risk of AR [26]. Patients with high ELISPOT counts, who received antibody induction therapy, had a significantly lower risk of AR than their counterparts who did not receive induction [27]. Post-transplant conversion to a negative ELISPOT assay occurred in 86% of patients who received induction therapy versus 35% of patients who did not [26, 28]. Contrary to previous reports, the large Clinical Trials in Organ Transplantation-01 (CTOT-01) observational study failed to identify an association between pre-transplant IFN-γ ELISPOT positivity and the incidence of AR or reduction in estimated glomerular filtration rate (eGFR) at 6- or 12-month in 176 primary kidney transplant recipients treated with heterogeneous immunosuppression [29]. Secondary analyses showed that, in patients not induced with thymoglobulin, positive pre-transplant-ELISPOT positivity was associated with lower 6- and 12-month eGFRs when compared with those patients with negative IFN-γ ELISPOT.

A study by the Fairchild's group showed that a significant subpopulation of alloreactive memory $CD28^{-}CD8^{+}$ T cells are not detected by the above-mentioned approaches [30]. Addition of IL-15 to the cultures rescues the proliferating capacity of these cells, making them detectable [31]. Since renal epithelial cells are able to produce IL-15 during inflammation, quantification of these cells might be relevant in quantifying the immune risk of kidney transplant recipients. Future investigations using ELISPOT should therefore consider the addition of IL-15 to cultures to assess this population of memory cells and their contribution to graft injury.

Panel of Reactive T-Cells
A T-cell reactivity index, or panel of reactive T-cells (PRT) has been proposed, utilizing the ELISPOT responses to common HLA antigens from a pool of donors that should be reflective of general organ donor pool. Similar to the panel-reactive antibody test for identifying individuals with elevated levels of anti-HLA antibodies, the PRT may identify patients at risk for post-transplantation cellular-mediated graft injury. In a small study of 30 kidney transplant patients, 6 of the 7 (86%) patients with AR were PRT-positive whereas only one had low PRT before transplantation [32]. Other small, retrospective studies reported an association between positive pre-transplant PRT and an increased risk of AR [33]. However, the utility of PRT in predicting graft outcomes and understanding the clinical variables associated with poor or protective outcomes has not yet been investigated in larger cohorts of kidney transplant patients.

Urine Biomarkers

As a direct product of kidneys and a noninvasive source of biomarkers, urine appears as an ideal compartment for immune monitoring of proteins and mRNA (Table 1).

C-X-C Motif Chemokine Ligand 9
Numerous studies have shown that urinary C-X-C motif chemokine ligand 9 (CXCL9), a chemokine produced by lymphocytes and graft parenchymal cells, is elevated in the setting of AR. In kidney transplant patients with acute increase in serum creatinine, urinary CXCL9 had a PPV of 75% and NPV of 92% for AR [34], particularly when concurrent urinary tract infections or BK virus positivity are simultaneously ruled in or out with cultures and molecular testing. Published data also showed significantly higher levels of urinary CXCL9 up to 30 days prior to the clinical recognition of allograft injury suggesting that this testing strategy could be used to detect subclinical injury (with a stable creatinine) that may develop over time into a rejection episode associated with graft function deterioration [34]. In the CTOT-09 trial, 21 kidney transplant recipients induced with thymoglobulin and maintained on tacrolimus (Tac), mycophenolate mofetil, and prednisone therapy, were randomized to withdraw Tac over 8 weeks or to remain on standard therapy, provided they had no DSA or signs of

rejection on a 6-month surveillance biopsy [35]. Of the 14 patients in the experimental arm, 5 developed AR episodes. In all of them, serial urinary CXCL9 measurements documented incipient rejection as early as 6 weeks prior to the event, whereas in patients whose safely withdrew Tac urinary CXCL9 was persistently negative [35].

Urinary RNA
Recognizing the increased allograft infiltration and subsequent urinary appearance of lymphocytes during AR episodes, pre-selected immune transcripts (T-cell/immune related) in urinary cell pellets measured by qPCR have been reported to noninvasively differentiate AR from non-AR (81–86Pe). One large, prospective study in 485 kidney transplant recipients showed that increased levels of CD3ε mRNA, IP-10 mRNA, and 18S rRNA in urinary cells appears to be diagnostic and prognostic of AR [36]. However, urine handling and storage for accurate transcript assessment needs much rigor and most of these studies on urinary RNA have emerged from a small number of specialized research groups, and their performance in clinic settings is uncertain at this time.

Molecular Biomarkers

Several groups have identified peripheral blood gene expression changes that have correlated with AR [37]. In theory, a peripheral blood signature could obviate the need for biopsy, while an allograft signature still necessitates it. Further, a peripheral blood signature for AR, or one that antedates the presence of AR, has the potential to serve as an immune monitoring tool. An interesting observation from simultaneous comparisons of blood and allograft transcriptomes of AR has been the consistent identification of lack of overlap between the gene profiles in blood and biopsy associated with AR [38]. Recently, a 17-gene qPCR-based assay in blood samples – the kidney solid organ response test (kSORT) was validated in a multicenter cohort including 438 renal transplant patients as a diagnostic and possibly predictive noninvasive test for AR [39]. A study in 75 consecutive kidney transplants who underwent 6-month protocol biopsies showed that the combined use of kSORT and IFN-g ELISPOT was able to noninvasively diagnose subclinical AR with an area under the curve >0.85% [40].

Other investigators [41] identified a peripheral blood gene signature that showed diagnostic accuracy for subclinical rejection, including borderline patients, early post-transplant. Although these approaches could potentially be used to guide therapeutic decisions in the absence of biopsy tissue, multicenter validation and commercialization are required (Table 1).

Cell Free-DNA

After graft injury, high amounts of graft-derived cell-free DNA (GcfDNA) are shed into the blood stream from dying cells. Immediately after engraftment, GcfDNA reaches high values (>5% of total GcfDNA), with a rapid decrease to values of <0.5% within 1 week. Living-related kidney transplant recipients show lower initial values than recipients of deceased organs, reflecting the absence of preservation injury. Episodes of rejection in kidney and heart transplant patients are accompanied by a significant increase of GcfDNA (>5-fold) above values in patients without complications, occurring earlier than clinical or biochemical hints to rejection [42]. However, available evidence suggests that elevations in cfDNA cannot distinguish AR from other causes of graft injury, and additional markers may need to be layered over this assay to improve specificity or accurately relay the need for allograft biopsy [43] (Table 1).

Future Directions and Conclusions

Although extensive efforts have been made in developing biomarkers to detect or predict kidney allograft rejection, few assays have moved from the research arena to a clinically implementable testing strategy that could guide therapeutic decision making.

The road to clinical biomarker implementation in transplantation is fraught with barriers that include high costs of commercializing assays for relatively small market sizes and a field that is being driven by physician-scientists rather than commercial entities. Improved partnering approaches among scientists, academic institutions, and various pharmaceutical and/or biotechnology companies will be required to commercialize biomarkers and make them available to transplant clinicians, particularly for assays with limited intellectual property rights, given that the methods are in the public domain (e.g., ELISAs for chemokines).

In addition, the relative lack of information on how to clinically use biomarkers remains problematic. The transplant field needs controlled trials, in which subjects are randomized to receive standard care or individualized care on the basis of pre-specified biomarker-directed strategies. End points should include graft outcomes, as well as immunosuppression-related toxicities and complications. The hypothesis to be tested is that biomarker-driven care will improve graft function/survival while decreasing the need for biopsies and limit infectious complications. Such controlled trials would provide the best evidence for how to use a given marker, but commercialized, FDA-approved biomarkers can also be broadly tested by the transplant community without hard evidence. Such real-world experience is important in generating hypothesis and driving improved trial design to address how an assay could be optimally used in a given setting.

These data will be instrumental for the kidney transplant community to shift from protocol-based immunosuppression to individualized treatment of patients driven by reliable biomarkers.

References

1 Meier-Kriesche HU, Kaplan B: Waiting time on dialysis as the strongest modifiable risk factor for renal transplant outcomes: a paired donor kidney analysis. Transplantation 2002;74:1377–1381.

2 Saundh BK, Baker R, Harris M, Welberry Smith MP, Cherukuri A, Hale A: Early BK polyomavirus (BKV) reactivation in donor kidney is a risk factor for development of BKV-associated nephropathy. J Infect Dis 2013;207:137–141.

3 Cravedi P, Heeger PS: Immunologic monitoring in transplantation revisited. Curr Opin Organ Transplant 2012;17:26–32.

4 Kissmeyer-Nielsen F, Olsen S, Petersen VP, Fjeldborg O: Hyperacute rejection of kidney allografts, associated with pre-existing humoral antibodies against donor cells. Lancet 1966;2:662–665.

5 Cravedi P, Heeger PS: Complement as a multifaceted modulator of kidney transplant injury. J Clin Invest 2014;124:2348–2354.

6 Loupy A, Lefaucheur C, Vernerey D, Prugger C, Duong van Huyen JP, Mooney N, Suberbielle C, Frémeaux-Bacchi V, Méjean A, Desgrandchamps F, et al: Complement-binding anti-HLA antibodies and kidney-allograft survival. N Engl J Med 2013;369:1215–1226.

7 Guidicelli G, Guerville F, Lepreux S, Wiebe C, Thaunat O, Dubois V, Visentin J, Bachelet T, Morelon E, Nickerson P, et al: Non-complement-binding de novo donor-specific anti-HLA antibodies and kidney allograft survival. J Am Soc Nephrol 2016;27:615–625.

8 Süsal C, Wettstein D, Döhler B, Morath C, Ruhenstroth A, Scherer S, Tran TH, Gombos P, Schemmer P, Wagner E, et al: Association of kidney graft loss with de novo produced donor-specific and non-donor-specific HLA antibodies detected by single antigen testing. Transplantation 2015;99:1976–1980.

9 Tait BD, Süsal C, Gebel HM, Nickerson PW, Zachary AA, Claas FH, Reed EF, Bray RA, Campbell P, Chapman JR, et al: Consensus guidelines on the testing and clinical management issues associated with HLA and non-HLA antibodies in transplantation. Transplantation 2013;95:19–47.

10 Wiebe C, Gareau AJ, Pochinco D, Gibson IW, Ho J, Birk PE, Blydt-Hansen T, Karpinski M, Goldberg A, Storsley L, et al: Evaluation of C1q status and titer of de novo donor-specific antibodies as predictors of allograft survival. Am J Transplant 2016;17:703–711.

11 Comoli P, Cioni M, Tagliamacco A, Quartuccio G, Innocente A, Fontana I, Trivelli A, Magnasco A, Nocco A, Klersy C, et al: Acquisition of C3d-binding activity by de novo donor-specific HLA antibodies correlates with graft loss in nonsensitized pediatric kidney recipients. Am J Transplant 2016;16:2106–2116.

12 Pinna D, Corti D, Jarrossay D, Sallusto F, Lanzavecchia A: Clonal dissection of the human memory B-cell repertoire following infection and vaccination. Eur J Immunol 2009;39:1260–1270.

13 Mulder A, Kardol MJ, Arn JS, Eijsink C, Franke ME, Schreuder GM, Haasnoot GW, Doxiadis II, Sachs DH, Smith DM, Claas FH: Human monoclonal HLA antibodies reveal interspecies crossreactive swine MHC class I epitopes relevant for xenotransplantation. Mol Immunol 2010;47:809–815.

14 Zachary AA, Kopchaliiska D, Montgomery RA, Leffell MS: HLA-specific B cells: I. A method for their detection, quantification, and isolation using HLA tetramers. Transplantation 2007;83:982–988.
15 Kopchaliiska D, Zachary AA, Montgomery RA, Leffell MS: Reconstitution of peripheral allospecific CD19+ B-cell subsets after B-lymphocyte depletion therapy in renal transplant patients. Transplantation 2009;87: 1394–1401.
16 Pape KA, Taylor JJ, Maul RW, Gearhart PJ, Jenkins MK: Different B cell populations mediate early and late memory during an endogenous immune response. Science 2011; 331:1203–1207.
17 Han M, Rogers JA, Lavingia B, Stastny P: Peripheral blood B cells producing donor-specific HLA antibodies in vitro. Hum Immunol 2009;70:29–34.
18 Snanoudj R, Claas FH, Heidt S, Legendre C, Chatenoud L, Candon S: Restricted specificity of peripheral alloreactive memory B cells in HLA-sensitized patients awaiting a kidney transplant. Kidney Int 2015;87:1230–1240.
19 Heidt S, Roelen DL, de Vaal YJ, Kester MG, Eijsink C, Thomas S, van Besouw NM, Volk HD, Weimar W, Claas FH, Mulder A: A NOVel ELISPOT assay to quantify HLA-specific B cells in HLA-immunized individuals. Am J Transplant 2012;12:1469–1478.
20 Lúcia M, Luque S, Crespo E, Melilli E, Cruzado JM, Martorell J, Jarque M, Gil-Vernet S, Manonelles A, Grinyó JM, Bestard O: Preformed circulating HLA-specific memory B cells predict high risk of humoral rejection in kidney transplantation. Kidney Int 2015;88: 874–887.
21 Bestard O, Cravedi P: Monitoring alloimmune response in kidney transplantation. J Nephrol 2016;30:187–200.
22 Woodland DL, Kohlmeier JE: Migration, maintenance and recall of memory T cells in peripheral tissues. Nat Rev Immunol 2009;9: 153–161.
23 Heeger PS, Greenspan NS, Kuhlenschmidt S, Dejelo C, Hricik DE, Schulak JA, Tary-Lehmann M: Pretransplant frequency of donor-specific, IFN-gamma-producing lymphocytes is a manifestation of immunologic memory and correlates with the risk of posttransplant rejection episodes. J Immunol 1999;163: 2267–2275.
24 Koscielska-Kasprzak K, Drulis-Fajdasz D, Kaminska D, Mazanowska O, Krajewska M, Gdowska W, Bieniecki W, Chudoba P, Polak W, Janczak D, et al: Pretransplantation cellular alloreactivity is predictive of acute graft rejection and 1-year graft function in kidney transplant recipients. Transplant Proc 2009; 41:3006–3008.
25 Bestard O, Cruzado JM, Mestre M, Caldés A, Bas J, Carrera M, Torras J, Rama I, Moreso F, Serón D, Grinyó JM: Achieving donor-specific hyporesponsiveness is associated with FOXP3+ regulatory T cell recruitment in human renal allograft infiltrates. J Immunol 2007;179:4901–4909.
26 Augustine JJ, Poggio ED, Heeger PS, Hricik DE: Preferential benefit of antibody induction therapy in kidney recipients with high pretransplant frequencies of donor-reactive interferon-gamma enzyme-linked immunosorbent spots. Transplantation 2008;86:529–534.
27 Cherkassky L, Lanning M, Lalli PN, Czerr J, Siegel H, Danziger-Isakov L, Srinivas T, Valujskikh A, Shoskes DA, Baldwin W, et al: Evaluation of alloreactivity in kidney transplant recipients treated with antithymocyte globulin versus IL-2 receptor blocker. Am J Transplant 2011;11:1388–1396.
28 Crespo E, Lucia M, Cruzado JM, Luque S, Melilli E, Manonelles A, Lloberas N, Torras J, Grinyó JM, Bestard O: Pre-transplant donor-specific T-cell alloreactivity is strongly associated with early acute cellular rejection in kidney transplant recipients not receiving T-cell depleting induction therapy. PLoS One 2015; 10:e0117618.
29 Hricik DE, Augustine J, Nickerson P, Formica RN, Poggio ED, Rush D, Newell KA, Goebel J, Gibson IW, Fairchild RL, et al: Interferon gamma ELISPOT testing as a risk-stratifying biomarker for kidney transplant injury: results from the CTOT-01 multicenter study. Am J Transplant 2015;15: 3166–3173.
30 Brzezińska A, Magalska A, Szybińska A, Sikora E: Proliferation and apoptosis of human CD8(+)CD28(+) and CD8(+)CD28(–) lymphocytes during aging. Exp Gerontol 2004; 39:539–544.

31 Traitanon O, Gorbachev A, Bechtel JJ, Keslar KS, Baldwin WM 3rd, Poggio ED, Fairchild RL: IL-15 induces alloreactive CD28(–) memory CD8 T cell proliferation and CTLA4-Ig resistant memory CD8 T cell activation. Am J Transplant 2014;14:1277–1289.
32 Poggio ED, Augustine JJ, Clemente M, Danzig JM, Volokh N, Zand MS, Hricik DE, Heeger PS: Pretransplant cellular alloimmunity as assessed by a panel of reactive T cells assay correlates with acute renal graft rejection. Transplantation 2007;83:847–852.
33 Uslu A, Nart A: Treatment of first acute rejection episode: systematic review of level I evidence. Transplant Proc 2011;43:841–846.
34 Hricik DE, Nickerson P, Formica RN, Poggio ED, Rush D, Newell KA, Goebel J, Gibson IW, Fairchild RL, Riggs M, et al: Multicenter validation of urinary CXCL9 as a risk-stratifying biomarker for kidney transplant injury. Am J Transplant 2013;13:2634–2644.
35 Hricik DE, Formica RN, Nickerson P, Rush D, Fairchild RL, Poggio ED, Gibson IW, Wiebe C, Tinckam K, Bunnapradist S, et al: Adverse outcomes of tacrolimus withdrawal in immune-quiescent kidney transplant recipients. J Am Soc Nephrol 2015;26:3114–3122.
36 Keslar KS, Lin M, Zmijewska AA, Sigdel TK, Tran TQ, Ma L, Bhasin M, Rao P, Ding R, Iklé DN, et al: Multicenter evaluation of a standardized protocol for noninvasive gene expression profiling. Am J Transplant 2013;13:1891–1897.
37 Sellarés J, Reeve J, Loupy A, Mengel M, Sis B, Skene A, de Freitas DG, Kreepala C, Hidalgo LG, Famulski KS, Halloran PF: Molecular diagnosis of antibody-mediated rejection in human kidney transplants. Am J Transplant 2013;13:971–983.
38 Sarwal M, Sigdel T: A common blood gene assay predates clinical and histological rejection in kidney and heart allografts. Clin Transpl 2013:241–247.
39 Roedder S, Sigdel T, Salomonis N, Hsieh S, Dai H, Bestard O, Metes D, Zeevi A, Gritsch A, Cheeseman J, et al: The kSORT assay to detect renal transplant patients at high risk for acute rejection: results of the multicenter AART study. PLoS Med 2014;11:e1001759.
40 Crespo E, Roedder S, Sigdel T, Hsieh SC, Luque S, Cruzado JM, Tran TQ, Grinyó JM, Sarwal MM, Bestard O: Molecular and functional noninvasive immune monitoring in the ESCAPE study for prediction of subclinical renal allograft rejection. Transplantation 2016, Epub ahead of print.
41 Menon M, Zhang W, Yi Z, Wei C, Woytovich C, Rosales I, Schroppel B, Chapman J, Nankivell B, O'Connell P, Smith RN, Colvin RB, Murphy B: A peripheral blood gene expression signature for subclinical acute rejection and associated long-term graft injury. Am J Transplant 2014;14:220–232.
42 Beck J, Oellerich M, Schulz U, Schauerte V, Reinhard L, Fuchs U, Knabbe C, Zittermann A, Olbricht C, Gummert JF, et al: Donor-derived cell-free DNA is a novel universal biomarker for allograft rejection in solid organ transplantation. Transplant Proc 2015;47:2400–2403.
43 Sigdel TK, Vitalone MJ, Tran TQ, Dai H, Hsieh SC, Salvatierra O, Sarwal MM: A rapid noninvasive assay for the detection of renal transplant injury. Transplantation 2013;96:97–101.

Paolo Cravedi, MD, PhD
Icahn School of Medicine at Mount Sinai
One Gustave Levy Place, Annenberg Building
New York, NY 10029 (USA)
E-Mail paolo.cravedi@mssm.edu

Author Index

Subject Index